The Right to Be Cold

THE RIGHT
TO BE COLD

One Woman's Fight to
Protect the Arctic and
Save the Planet from
Climate Change

Sheila Watt-Cloutier

Foreword by Bill McKibben

University of Minnesota Press
Minneapolis
London

The University of Minnesota Press gratefully acknowledges the generous assistance provided for the publication of this book by the Hamilton P. Traub University Press Fund.

Originally published in 2015 by Penguin Canada Books, Inc.

Published in 2018 in the United States by the University of Minnesota Press

Published by the University of Minnesota Press
111 Third Avenue South, Suite 290
Minneapolis, MN 55401–2520
http://www.upress.umn.edu

Printed in the United States of America on acid-free paper

The University of Minnesota is an equal-opportunity educator and employer.

24   23   22   21   20   19   18   10   9   8   7   6   5   4   3   2   1

Library of Congress Cataloging-in-Publication Data
Names: Watt-Cloutier, Sheila, author.
Title: The right to be cold : one woman's fight to protect the Arctic and save the planet from climate change / Sheila Watt-Cloutier ; foreword by Bill McKibben.
Description: Minneapolis, MN : University of Minnesota Press, 2018. Includes index.
Identifiers: LCCN 2017056053| ISBN 978-1-5179-0497-5 (pb)
Subjects: LCSH: Watt-Cloutier, Sheila. | Environmentalists–Canada–Biography. | Human rights workers–Canada–Biography. | Inuit women–Canada–Biography. | Inuit–Canada–Social conditions. | Environmental protection–Arctic regions. | Climatic changes–Arctic regions. | Arctic regions–Environmental conditions. | BISAC: BIOGRAPHY & AUTOBIOGRAPHY / Personal Memoirs. | SCIENCE / Earth Sciences / Meteorology & Climatology.
Classification: LCC GE56.W28 A3 2018 | DDC 363.70092 [B]–dc23
LC record available at https://lccn.loc.gov/2017056053

For my late grandmother—my Atikuluk—Jeannie
and my late mother, Daisy
For my children, Sylvia and Eric
For my grandsons, Mister Lee and Inuapik
For the love of my Inuit culture

In memory of Terry Fenge

In a time lacking in truth and certainty and filled with anguish and despair, no woman should be shamefaced in attempting to give back to the world, through her work, a portion of its lost heart.

Louise Bogan, American poet

CONTENTS

FOREWORD

Bill McKibben

DUE TO THE PECULIAR PHYSICS of climate change, the Arctic has warmed much faster than any region of the globe. Humans trap heat in the atmosphere with the cloud of carbon dioxide we've created by burning coal and gas and oil; that extra warmth is the heat equivalent of 400,000 Hiroshima bombs daily. That helps you understand how half the sea ice in the Arctic is now gone: a meters-thick, millennia-old shield of ice has quickly thawed. Viewed from a satellite, the Arctic looks utterly different from a few decades ago—much less white, much more blue. And of course the process builds on itself: that blue water absorbs the rays of the sun that the white ice used to deflect. This "Arctic amplification" guarantees that more warming is ahead for this land so long defined by the cold.

Sheila Watt-Cloutier is just old enough to remember what might be called the old Arctic. Born in the dark early winter of 1953 in the town of Kuujjuaq in northern Quebec, she didn't know the pristine Arctic of her forebears; the Hudson's Bay Company had come through generations before, and World War II had brought an American air base. But the basic outlines of Inuit culture were still fully intact—and so was the cold. "As the youngest child of four on our family hunting and ice-fishing trips, I would be snuggled into warm blankets and fur in a box tied safely on top of the *qamutiik,* the dogsled. I would view the vast expanses of Arctic sky and feel the crunching of the snow and the ice below me as our dogs . . . carried us safely across the frozen land" (xvii). She spent her first ten years in that world, swaddled in a loving family and community, protected above all by her grandmother, who was a living link to the older days. (The description of how her grandmother would salute young hunters after their first successes by acting out the part of the animals is marvelously told.) Watt-Cloutier spent her first years "bonded with the ice and snow" (xviii).

But then the world did its best to break all those bonds, one after another. Watt-Cloutier was sent south to school at the age of ten. At this distance it's hard to judge the intent of this policy, but benevolent or not it was a crushingly hard burden on a young girl, separated from all that she knew and tossed into a new culture. She did well academically but describes that "our achievements at school came at the cost of our Inuit knowledge and skills" (33). In just two years "we lost a remarkable amount of our mother tongue. . . . During the years we would spend away from home, other girls our age were learning how to prepare country food, embroider, and sew. They were mastering the cultural skills of working with caribou hide, sealskin, and goose down" (33). As the school years continued, the returns home became painful for other reasons: it became clear that the imposed rush to modernity was taking a huge toll on communities. Alcohol spread widely, and so did a kind of hopeless despair. I guarantee you won't read the accounts of the day the Royal Canadian Mounted Police shot teams of sled dogs in front of their owners without feeling a certain rage. You shouldn't read the stories of Watt-Cloutier's experiences as a medical interpreter without grieving for all the people around the world who have seen their lives, and their bodies, turned upside down by the imperialism of a consumer society.

And of course in the Arctic this colonial reach carried an even deeper burden: the waves of pollution that increasingly washed up from the south, transforming the landscape and with it the life ways of an entire people. First came the persistent organic pollutants, or POPs. No one was using DDT in the far north, but it bio-accumulated up the marine food chain to the point where the breast milk of Inuit women was the most contaminated on the planet. This was Watt-Cloutier's first great battle on behalf of her people; she and her colleagues worked the endless UN bureaucracy and the wearisome chain of meetings and conferences until they won the changes that were necessary to start reining in this toxic burden.

In a sense, that fight was the preparation for the even larger battle against climate change—and the even more powerful forces of the fossil fuel industry. As waves of warm air pushed up from the south, the Arctic was changing fast. Places that had never seen a thunderstorm now saw them regularly. Fire spread degrees of latitude north into the Siberian taiga. Really, *everything* changed. "In Nunavut," Watt-Cloutier writes, "people reported that since the ice was becoming so unreliable in the warmer springs, they could no longer travel out to the islands to collect eggs, geese, and seal. By the time they could get to the islands by boat, the eggs were too old to harvest" (188). Hunters could no longer cross bays, since the ice was too weak—following the ocean shoreline made the hunt much longer. Without the protection that shore ice provides from waves, coastal communities were battered by erosion, and entire Alaskan villages had to face the prospect of moving away from the places they had inhabited since time began. The deep, dense snow needed for building igloos became harder to find; hunters had to carry tents now, bulky burdens that didn't provide much insulation or much protection from polar bears. Even the animals were changing. "The seals often appeared to have less fat, which meant they sank deeper in the water, making them tougher to catch" (189). Those who depended on reindeer or caribou, like the Sami or the G'wichin, reported that warm weather followed by a freeze would put a layer of ice atop the lichen and grass, making it impossible for the animals to break through and feed.

And here's the thing. Each of these changes didn't just make life harder in practical terms. It also undercut the basis of the culture that had thrived for so many millennia: "It was becoming increasingly difficult for us to pass on our traditional knowledge, survival skills, and cultural richness to our children" (202). As the heat (caused, we must never forget, by a series of choices in the rich and settled world) wipes away the ice, so too it is wiping away much of the remarkable civilization of the north.

BUT IT ISN'T GOING WITHOUT A FIGHT, and that fight helps not only in the larger battle but also to revivify that great civilization.

Sheila Watt-Cloutier is a true hero. She has stayed calm, controlled, and modulated for decades now as she has led the fight against pollutants and against climate change—led the fight for her Inuit brothers and sisters across the far north. Part of that is her temperament; she describes in moving terms how Inuit learn early to be silent on the hunt, and how that carries over to other aspects of their life, a useful discipline that is breaking down under the pressures of modern, southern life. But she's also had to be controlled in order to be taken seriously: the councils of Very Important People are programmed to treat those from frontline communities in paternalistic and sentimental ways. Too often these people have had to suppress a certain amount of righteous anger to be taken even half-seriously.

But the upsurge of Indigenous environmentalism is one of the great triumphs of recent years. The emergence of many leaders like Watt-Cloutier, and of strong and unified communities behind them, has reshaped the struggle from Standing Rock to the Australian outback. And as the rest of us follow this lead, there is no reason for any of the rest of us to hold back in any way. To support the struggles of the Inuit (or of the Native Americans threatened by pipelines, or the Pacific Islanders threatened by rising seas, or of Bangladeshis whose crop land is submerging, or African Americans living amid belching and dangerous refineries) is to stand up against a threat that will eventually take down even the richest and safest—it's ultimately in everyone's self-interest. But it's also standing up for justice, which in this context is an even more important word than climate. If you have spent your life pouring carbon into the atmosphere, your job is not to feel guilty—that doesn't help. And your job is not, primarily, to worry about your lightbulbs or your lifestyle; at this late date those changes, though important, are not going to protect the Arctic or anything else. At this late date our job is to build movements, ones powerful enough to force the policy changes that give us our only hope of catching up with physics.

Those movements—and the careful behind-the-scenes diplomacy that Watt-Cloutier describes—have forced some action. The Paris climate accords, signed in 2016, were a high-water mark of those efforts. The United States under Donald Trump has already announced it won't follow them—a gross abdication of responsibility that should be an abiding source of shame for any American. But even if we, and everyone else, had followed the letter of those agreements, the planet would still have warmed 3.5 degrees Celsius, 6 or 7 degrees Fahrenheit. And remember, the top of the world will warm far more than that.

So the call is obvious. We need to fight back hard, for much faster progress than envisioned in Paris. We know what's required: the technology of sun and windpower has become so good and so cheap in the past few years that it's entirely possible to imagine the rapid conversion to 100 percent renewable energy. And we know what's in the way: the fossil fuel industry and its political allies, who want to extend their business plan a few more decades, even at the cost of breaking the planet. What we don't know is what's in our hearts: do we have the courage and commitment to build the nonviolent movements that could make a difference? Reading this book—the story of a quiet young woman who rose to lead her people in a desperate battle—should give all of us the inspiration we need, whether it's to go to jail blocking pipelines or to run for Congress battling oil companies. We owe Sheila Watt-Cloutier an immense debt.

INTRODUCTION

THE WORLD I WAS BORN INTO has changed forever.

For the first ten years of my life, I traveled only by dog team. As the youngest child of four on our family hunting and ice-fishing trips, I would be snuggled into warm blankets and fur in a box tied safely on top of the *qamutiik*, the dogsled. I would view the vast expanses of Arctic sky and feel the crunching of the snow and the ice below me as our dogs, led by my brothers, Charlie and Elijah, carried us safely across the frozen land. I remember just as vividly the Arctic summer scenes that slipped by as I sat in the canoe on the way to our hunting and fishing grounds. The world was blue and white and rocky, and defined by the things that had an immediate bearing on us— the people who helped and cared for us, the dogs that gave us their strength, the water and land that nurtured us. The Arctic may seem cold and dark to those who don't know it well, but for us a day of hunting or fishing brought the most succulent, nutritious food. Then there would be the intense joy as we gathered together as family and friends, sharing and partaking of the same animal in a communal meal. To live in a boundless landscape and a close-knit culture in which everything matters and everything is connected is a kind of magic. Like generations of Inuit, I bonded with the ice and snow.

Those idyllic moments of my childhood seem very far away these days. Today, while dog teams, *qajaq*s (kayaks), and canoes are at times still used to move out onto the Arctic land and water, snow machines are more common than dogs, and the hum of fast-moving powerboats is now heard on Arctic waters. All of our communities now have airports, medical clinics, and schools, with some having hospitals, television stations, daycares, and colleges. Our people still hunt and fish, sew and bead, but they are also nurses, lawyers, teachers, business people, and politicians. The Arctic is a different place than it was when I was a child. And while many of the changes are positive, the journey into the modern world was not an easy one—and it has left its scars.

In a sense, Inuit of my generation have lived in both the ice age and the space age. The modern world arrived slowly in some places in the world, and quickly in others. But in the Arctic, it appeared in a single generation. Like everyone I grew up with, I have seen ancient traditions give way to southern habits. I have seen communities broken apart or transformed dramatically by government policies. I have seen Inuit traditional wisdom supplanted by southern programs and institutions. And most shockingly, like all my fellow Inuit, I have seen what seemed permanent begin to melt away.

The Arctic ice and snow, the frozen terrain that Inuit life has depended on for millennia, is now diminishing in front of our eyes.

We are all accustomed to the dire metaphors used to evoke the havoc of climate change, but in many parts of the Arctic the metaphors have already become a very literal reality. For a number of reasons, the planet warms several times faster at the poles. While climate experts warn that an increase of two degrees in the global average temperature is the threshold of disaster, in the Arctic we have already seen nearly *double* that. As the permafrost melts, roads and airport runways buckle. Homes and buildings along the coast sink into the ground and fall into the sea. The natural ice cellars that are used for food storage are no longer cold. Glaciers are melting so fast that they now create dangerous torrents. The world becomes focused and horrified only by haunting images of polar bears struggling to find ice, but hunters too are finding that the once reliable ice can be deadly. The land that is such an important part of our spirit, our culture, and our physical and economic well-being is becoming an often unpredictable and precarious place for us.

I worry not just as an Inuk, but as a grandmother too. In our culture, hunting has taught us to value patience, endurance, courage, and good judgment. The hunter embodies calm, respectfulness, caring for others. *Silatuniq* is the Inuktitut word for wisdom—and much of it is taught through the experiential observation of the hunt. The Arctic is not an easy place to stay alive if one has not mastered the life skills passed down from generation to generation. Mistakes can be fatal. But every challenge teaches a lesson, not only about the techniques of thriving in a cold world but also about developing the character that can be counted upon to stand up to those challenges. It is the wisdom of our hunters and elders that allowed us not only to live but also to thrive. As you are taught how to read the weather and ice conditions and how to become a great hunter or a great seamstress, you learn to become focused and meticulous, for your family depends on these skills for survival. This is the wisdom our hunters and elders have shared with our children for generations, and this holistic approach to learning is an essential part of Inuit culture. My hope for my grandsons is that they will inherit a good part of this culture. But this important traditional knowledge has begun to lose its value as a result of the dramatic changes to our environment. This wisdom, which comes from a hunting culture dependent on the ice and snow, is as threatened as the ice itself.

The environment and climate I grew up in was indeed rich in lessons, and not just those that build character or help us on a hunt. Our intense affinity with the land and with wildlife taught us how to live in harmony with the natural world. Our traditional hunting and fishing practices do not destroy habitat. Nor do our practices deplete animal populations, or create waste. We use every part of the animals that we harvest. In other words, for thousands of years, Inuit have lived sustainably in our environment. We have been stewards of the land. All this wisdom, too, is threatened by the changing climate. That is to say, if we allow the Arctic to melt, we lose more than the planet that has nurtured us for all of human history. We lose the wisdom required for us to sustain it.

And when I say *we*, I do not mean only Inuit. It is true, we are already among the first to be devastated by climate change, but we are not the only ones.

Everything is connected through our common atmosphere, not to mention our common spirit and humanity. What affects one affects us all. The Arctic, after all, is the cooling system, "the air conditioner," if you will, for the entire planet. As its ice and snow disappear, the globe's temperatures rise faster and erratic weather becomes more frequent. This results in drought, floods, tornadoes, and more intense hurricanes. Sea levels around the world rise, and small islands from the Caribbean to Florida to the South China Sea slip into the ocean. From the farmers in Australia to the fishermen in the Gulf of Mexico or the homeowners of New Orleans, the devastation escalates. The future of Inuit is the future of the rest of the world—our home is a barometer for what is happening to our entire planet.

For years, our Inuit communities have been trying to draw the world's attention to the environmental disaster we are witnessing. But the battle to save our lands and our way of life has met many challenges. The industrialized nations' fears of job losses and short-term economic decline prevent many of them from making significant reductions to the production of greenhouse gases and other pollution. Developing countries argue that they should not be slowed down by environmental restrictions that might hinder their industrial growth. As astonishing as it is to those of us who are watching our inheritance begin to melt away, there are those who see the destruction of this irreplaceable world as an opportunity. Commercial flights arrive daily in the North filled with those from mining, oil, and gas companies who are eager to exploit the riches below the melting ice.

This is, of course, short-term thinking. More and more studies suggest that climate change will be the economic ruin of our world—it will cost us much more to address the damage done by climate change and to adapt to a devastated environment (to the tune of trillions of dollars by mid-century) than we are ever going to save by using "cheap" sources of energy or by allowing our industries and manufacturers to pollute.

While the economic argument may eventually move government and industry to serious action, the debate about the numbers, not to mention the science, allows industry and governments to call for more delays, to drag their feet about reducing greenhouse gas emissions or developing sustainable energy. There is, however, another way to fight for the protection of our planet: to demand that the global community recognize that the well-being of our environment is in itself a fundamental human right. Without a stable, safe climate, people cannot exercise their economic, social, or cultural rights. For Inuit, as for all of us, this is what I call *the right to be cold*. And this is what I have been fighting for over the last twenty years of my life's work.

In 2005 I was part of a team of environmental lawyers, Inuit women, and Inuit hunters that filed a petition with the Inter-American Commission on Human Rights, asking that the protection from climate change be recognized as a fundamental human right by member countries of the commission. My role in this landmark effort was to tell the world about what was happening to Inuit and our culture as a result of rising temperatures and melting ice. I was, in other words, putting the human face on the climate change debate. (And as a result of this work I was co-nominated with Al Gore for the 2007 Nobel Peace Prize.) Our efforts were met with interest, enthusiasm, and support from people all over the globe. Since that time, other initiatives have led to studies, statements, and resolutions from the United Nations Human Rights Council, the UN High Commissioner for Human Rights, and the Inter-American Commission on Human Rights that recognize the link between human rights and climate change. But there is much more to be done.

I believe the power of a human rights–based approach is that it moves the discussion out of the realm of dry economic and technical debate. A human rights–based approach takes the path of principle, showing us that fundamental change is not just sound policy but also an *ethical* imperative. It refocuses the debate on humanity rather than solely on economics. And in doing so, it is one of the most important tools we have in our ongoing battle to save our environment and one of our best hopes to save the planet. In other words, I believe the campaigns to link climate change to human rights protection—efforts that acknowledge our shared humanity and our shared future—are the most effective way to bring about lasting change.

It may be surprising to some that while fighting for the right to be cold has arguably been the hallmark of my life's work, I do not consider myself an environmental activist. I came to be involved in environmental issues through my global work as an elected leader advocating for my circumpolar Inuit community. An early job with the medical clinic in my hometown of Kuujjuaq gave me an intimate look at the challenges our people were facing. In the following years, my work with both the Kativik School Board and the Nunavik Education Task Force provided more insight into the struggles and the barriers that our youth and our future were facing. But when I was elected as the corporate secretary of the Makivik Corporation and the president of the Inuit Circumpolar Council (ICC) Canada, I joined the international struggle to eliminate the persistent organic pollutants that were finding their way into Arctic waters, Inuit food sources, and Inuit bodies. From there, as an elected Inuk official and chair of ICC International, I was launched into international climate change politics. My work was global, but the protection of my Arctic home and my Inuit community always drove my efforts. While climate change became the focus of much of my work, it was clear to me that a holistic approach must be taken to heal the wounds that affect Inuit communities—historical traumas; current spiritual, social, health, and economic problems; *and* the environmental assaults on our way of life. Our challenges cannot be "siloed" or looked at in isolation if we want to rise above them. So the story of *The Right to Be Cold* is also, in part, the story of Inuit history and contemporary Inuit life, through my lens.

As an Inuk woman, a mother, and a grandmother who feels blessed to have been born into this remarkable culture, I want to offer a human story from this unique vantage point. In essence, the goal of my book is to share with the world the parallels I see between the safeguarding of the Arctic and the survival of my Inuit culture. Writing *The Right to Be Cold* is also my way of giving back to the people and the culture that have served not only as my grounding foundation but also as the very anchor of my spirit as I was propelled into the rumble-tumble world of international politics.

My own personal journey of struggle and loss mirrors that of many people in our Inuit communities, as well as in other Indigenous and vulnerable communities around the world who have been marginalized and who fight to reclaim their sense of control and find their ground. It is so important to me that my grandsons understand their *anaanatsiaq*'s (grandmother's) life's work, not in terms of the awards that sit in her living room, but rather in the meaning behind each one of these accolades. To become overidentified and consumed with dark thoughts created, in large part, by our historical traumas leads all too often to our youth taking their own lives rather than embracing their lives. From the insights gained from my own journey, I attempt to put these historical events into perspective so those who struggle may better understand that much of our fear originates from the traumas of our forefathers. While the profound and personal pain we feel today does stem from our own childhoods, I feel it is important to understand that it is compounded by what we have inherited from our intergenerational histories. Therefore, not taking this received trauma solely as personal is vital to healing. My maternal instincts also tell me that it is extremely important that the younger generation see beyond the material recognition and grasp the meaning of success: achievements reflect an acceptance of the human condition, with all its challenges, and they mark the human journey, which requires us to show up and focus, commit to our passions, persevere, and endure the moments of struggle and loss in order to overcome them and transform our lives. I was like everyone else. I *am* like everyone else. I did not set out to change the world or become a Nobel Peace Prize nominee.

I hope that this book also helps to correct the preconceived notions of the Arctic and Inuit that many people hold. The Arctic is not a frozen wasteland. Its ice and snow are teeming with life—not just marine and animal life, but human life: men, women, and children; families and communities. Every time I fly home from far-flung places I can't help but smile as a northern town comes into focus against the blue-tinged vastness of ice, snow, and sky. If one focuses only on the geographical location, it can be hard to believe that these small, isolated communities dotting the far reaches of the Arctic can be "warm" and have a palpable welcoming energy, but they certainly do. They also have a culture rich in traditional wisdom, collective spirit, and technological and artistic skill. The Arctic is much more than polar bears and seals.

But I also hope that this book inspires others to take up the cause of climate change. What is happening today in the Arctic is the future of the rest of the world. In one lifetime, we Inuit have seen our physical world transform, the very ground beneath our feet shift dramatically. We have witnessed our once stable weather become erratic, unpredictable, and dangerous. We've watched as the wildlife that sustain us physically and spiritually struggle for their own health and survival. All of this has happened in a matter of decades, and the dire changes only promise to come faster if the global community does nothing to slow climate change.

As we head into stormier seas, we must ask ourselves, *If we cannot save the frozen Arctic, how can we hope to save the rest of the world?*

1

AN EARLY CHILDHOOD
OF ICE AND SNOW

DESPITE ALL THE CHALLENGES that were to come for me and my Arctic home, my early years were happy ones, filled with love, security, and a profound sense of belonging. I grew up in the presence of strong, independent women and in the arms of a community that was in the midst of great changes, but was still close-knit, inclusive, and supportive.

When I was born in 1953, my small family lived in Old Fort Chimo, a tiny community in northern Quebec, in the region that would eventually become known as Nunavik. Fort Chimo was a mispronunciation of *saimuuq*, which means handshake, and was the name the Hudson's Bay Company (HBC) gave the community. In Inuktitut, the place had been called Kuujjuaq, meaning "great river," for its location along the Koksoak River (a name given to it by Moravian missionaries, probably a mispronunciation of Kuujjuaq).

Kuujjuaq sits just below the treeline, and the rolling, hilly terrain is dotted with tamarack and black spruce trees. During the short summer months, cloudberries, blueberries, arctic cranberries, and black crowberries grow among the green leaves and tundra. Fluffy white cottongrass and deep-pink fireweed wave in the breeze, and bluebells, green mosses, and grey lichens spread out across the land. In the winter, the landscape is transformed into a brilliant vista of ice and snow that stretches under the vast expanse of the blue Arctic sky.

Nestled in this beautiful open landscape was Old Fort Chimo, a former Hudson's Bay Company trading post, which had opened in the 1800s or early 1900s. The community had expanded between 1948 and 1950, but there were still only a few family homes when I was a child—small painted plywood structures heated with oil or wood stoves, and without running water or electricity. The Hudson's Bay store and its storage warehouses, the Roman Catholic church, a nursing station, and a weather station, all painted plywood as well, made up the rest of the village. There were no cars in Old Fort Chimo that I can recall. Occasionally, Hudson's Bay employees or government officials would ride through town in early versions of the Bombardier snowmobile—boxy enclosed vehicles on rubber treads. But for Inuit, our transportation was *qimutsiit* (dog teams) or our own feet.

The Inuit population in Old Fort Chimo amounted to only a few dozen people in those years. Most families still lived in outpost camps, up to several days' travel away, where they hunted, fished and trapped. These families would come to the settlements only to barter their furs for food and ammunition at the HBC store. But there were no hunters in my immediate family in the early days, so my grandmother stayed at the HBC post.

We lived in Old Fort Chimo until I was four years old, when we moved to New Fort Chimo across the river. New Fort Chimo was home to a former American air force base called Crystal I, which had been built during the Second World

War. The base was part of the Crimson Route of airstrips that connected Great Britain with North America as a means for U.S. forces to support the Royal Air Force. In total, four bases were built in North America, and three in Greenland. New Fort Chimo was a location of strategic importance to the Americans, serving as an in-between on the way to Europe. In Fort Chimo, Frobisher Bay (Iqaluit), and Padloping Island, our Inuit communities served as stepping stones.

Many years after the war ended and the Americans had left, the landing strip in New Fort Chimo was still used by the planes that brought in supplies from the South. Traveling by canoe in the summer and by dog team in the winter, our men would pick up what they needed from the former American base. We had gone there from time to time when I was a child to collect our mail and run errands. But it was inconvenient for everyone to cross the river for the post and supplies, so eventually all the services, including the Anglican and Catholic missions and the nursing station, relocated to New Fort Chimo, and a school was built. The families of Old Fort Chimo followed. By the time my family moved, the only thing that remained across the river was the Hudson's Bay store, and my uncle's family, as he worked there. Eventually, the store and my uncle would move as well.

New Fort Chimo grew rapidly over the next decade, as more and more Inuit families relocated there. The abandoned military buildings were used to house services and families in the early years, but were eventually replaced by government-built homes and offices. Evidence of the U.S. presence, however, remained. The Americans had left behind hundreds of barrels full of tar, which I assume were used to build the airstrip, and which weren't removed until the town of Kuujjuaq undertook the task when I was an adult. Along with the barrels dotting the landscape, the earth was scored with the tracks of American tanks and trucks—marks that remain on the land to this day. As New Chimo grew, cars became more and more common. By the time I was in my teens, Kuujjuaq would be a town of hundreds of people. Today, its population is over two thousand.

While each passing year in New Chimo would make our permanent Inuit community larger, in Old Chimo my family had been somewhat isolated from our extended family and the greater Inuit community. That was largely because our family was different from both during my childhood, and during my mother's.

Sometime around 1920, my mother's mother had met a Scotsman named William Watt. Watt had come north to work as a Hudson's Bay Company apprentice clerk. During his time in Fort Chimo, he and my grandmother had three children. In the summer of 1926, however, William Watt left Fort Chimo and his Inuit family. My mother, Daisy, was only five years old. Her sister, Penina, was a toddler. My uncle Johnny was born after his father's departure, in November.

In those days, it was not the norm for *qallunaat* men who worked for the HBC to stay and marry the Inuit women with whom they had fathered children. (*Qallunaat* is the Inuktitut plural term for white people. It's derived from *qallunaq*, which describes the bones on which the eyebrows sit, which protrude more on white people than on Inuit.) When it was time for the managers to transfer to another post, they did so, most often leaving the mothers and their children behind.

It does appear, however, that my grandfather asked my grandmother to come with him. He wrote a letter to my grandmother after he left, telling her that he thought she would like the place where he now lived. But as much as my grandmother didn't want to lose the man she surely loved, and the father of her three children, she couldn't imagine living in the southern world. She felt that she had to stay behind, while my grandfather moved on. It was clearly a difficult choice. I was once told that my grandmother said to William Watt, "If you must go, if you must marry, don't marry an Inuk or Indian woman so it won't hurt me as much." She would later hear that he had married and had returned north to another community with his white wife and son. But he never contacted her, nor did he ever mention his Inuit family when interviewed decades later about his northern experiences.

While my grandfather's move might not have been unusual for the time, it was devastating nevertheless. My mother was abandoned by a father she had loved; my uncle would never know his father. And my grandmother was faced with a heart-rending dilemma. She could no longer feed and care for both of her little girls, and she was expecting another child. The only way she could ensure the survival of all her children was to give two-year-old Penina to another family in the community, the Shipaluks, to raise. My grandmother was a gentle, loving mother and grandmother. It's hard to imagine how difficult the decision must have been for her. And while Penina saw her mother and her siblings all the time as she grew up, and the Shipaluk family cared for her well, she struggled her entire life with the painful knowledge that her family had been unable to keep her.

Even with Penina in the care of another family, my grandmother was in a difficult position. William Watt may have provided some monetary support for a time (and when my uncle Johnny became a young man, he would ask the Hudson's Bay Company to hire him), but it was not enough. And there were no hunters in the household. While my grandmother did manage, with very little hunting gear, to hunt easy prey like ptarmigan, she could do this only close by the community. She couldn't leave her children to join the men who went out onto the land for extended periods. Her only option was to stay in Fort Chimo and work for the HBC store, doing domestic labor. By the time my mother was ten, she was helping to support the family, working alongside her mother at the store, where she would get some food as pay. When Johnny became a teenager, he was taught to hunt by older relatives, and with his earnings from the HBC store he was able to contribute to the family as well.

As a single woman, my grandmother had to work harder than most to provide for her family. This was an experience my own mother would share.

My mother had three children—Charlie, then Bridget six years later, and me, Sheila (Siila), four years after Bridget—from three different relationships. Each one of these *qallunaat* men would leave her. The last was my father. She and my grandmother also adopted a twelve-year-old boy, Elijah, before I was born. My mother had to work hard to support us all.

But my mother had an advantage that my grandmother had not had. When she was young, my mother had been taught English at the Catholic mission by a missionary named Umikutaak, meaning "long beard." She was one of the rare Inuit in that era who was proficient enough in English to be considered bilingual. By the time I was born, this achievement had landed her a job as an interpreter for the staff at the nursing station in Fort Chimo and for the health teams that traveled into Nunavik communities by plane.

Like her mother before her, my mother had to keep her family in the HBC post of Old Fort Chimo while the other families headed out to the hunting and fishing grounds. But we never felt we lacked for anything. Both my grandmother and my mother played the dual role of mother and father to us, and we felt secure in their care. Indeed, my mother and my grandmother were my mentors. They were remarkably resourceful women. I never saw them in any kind of weak position, nor did they ever complain, despite the fact that their lives were without luxury or ease. Instead, they showed us dignity and integrity, strength and perseverance. They not only survived as single mothers but also thrived.

My mother's resourcefulness was evident in her decision to build her own home when I was about seven.

When we first arrived in New Fort Chimo, we moved into a section of an old American building. It was long and covered with black tarpaper, and had a cooking and living area but no real bathroom. I don't know if another family lived at the other end of the building, but it never felt like home, perhaps because I knew it was temporary. Eventually, our family was given one of the rigid frame houses that the government was providing to Inuit families at the time. These were tiny, cramped structures that looked a bit like covered tent frames. (The more square government-issued homes were called "matchbox" homes because of their size.) The government house had electricity but no running water. A half wall separated the sleeping quarters from the kitchen and living spaces, while a small area was sectioned off for the toilet—a honey bucket that we emptied every day. Even though the house was tiny—there were six of us living there, after all—it felt cozy and familiar, much closer to our Old Fort Chimo home than the dark, drab American building.

But after several years in the government home, my mother took the unusual step of building her own house, with the help of my brother Elijah. They managed to acquire wood from a dismantled power generator building in Old Fort Chimo, and

Elijah and a few others brought it across the river by canoe. My mother purchased the rest of the materials they needed.

Elijah had gained a lot of construction experience working on various buildings that had been erected in New Chimo over the years. He was guided by Uncle Johnny, who was a great carpenter and who had also built his own home. (Most Inuit men in those days had become jacks of all trades, applying the ingenuity and architectural and engineering abilities they'd mastered in traditional life to learning the new skills they needed to build houses, fix furnaces, and repair snowmobiles.) My mother worked alongside Elijah on the house, often long into the evening, with the help of one or two men from the federal government construction crew after hours. When they finished, we had a home with a kitchen separate from the living room, as well as three small bedrooms, a bathroom, and a porch. The house had an oil stove in the kitchen and an oil space heater in the living room. And while we still didn't have running water (my mother wouldn't have this luxury until the year I turned twenty), we could heat water on the stove for baths and for laundry and dishes.

Our move into the house my mother built filled the whole family with a sense of pride and accomplishment. We knew it was rare for Inuit to build their own homes—much less an Inuk woman. It spoke to my mother's feistiness and her confidence.

Indeed, my mother was ahead of her time. She was an excellent provider, making sure her children always had everything they needed, plus a few luxuries, like trendy clothes and the modern things that teenagers everywhere were into. (When I was a teenager, she bought me a battery-operated record player because she knew I loved music and dance so much.) She was also fiercely independent and not at all shy about giving her opinion. And while the adults in our community all looked out for everyone's children, my mother was perhaps more forward than many in shooing a young child indoors when she felt he was out too late at night, or telling a patient she visited with the medical team that he needed to take more care with his hygiene. She was an extrovert, through and through.

She could also be stubborn and had difficulty showing affection to her children. Although my mother had a lively sense of humor and could be playful and fun with kids, she didn't reveal her warm side too often to us. (But when the grandchildren came along she was consistently affectionate with them.) Her reserve was especially evident with me. On rare occasions she would share memories of the first two men in her life, the fathers of my sister and brother, but she remained guarded about my father. It was clear that his abandonment of her was especially painful, and that pain seemed to have built a little wall around her heart when it came to me.

But any lack of affection from my mother was made up for by the gentle warmth of my grandmother. Every day when my mother went to work, my grandmother stayed home, gracing us with her calm and loving presence. I awoke each morning to the sounds of the radio. As the voices from the Greenland station and the CBC North Radio broadcast drifted into the bedroom, I could hear her making tea and breakfast for me, usually oatmeal and bannock. During the rest of the day, we visited friends and family together or had people over for tea and bannock. The Inuit community of my youth was not generally physically demonstrative between adults, but they always gave babies and small children plenty of hugs, cuddles, and Inuk *kuniks* (kisses), which are more like little sniffs on the cheek. And when children get a little bigger, we use words, looks, and subtle gestures to show affection. We make a little murmur at the back of our throats, a sort of soft *uummpp* that is like a verbal caress. The hours I spent with my grandmother and other adults of the community were always filled with that gentle sound.

As a small child in Old Fort Chimo, I would perch upon the window ledge in the kitchen each evening, watching for my mother to come home from work. But what I remember most about the nights was my grandmother telling me stories or reading the Bible to me as I fell asleep. When I was a little older, I used my grandmother's cherished hymn book, which was written in Inuktitut syllabics, or Qaniujaaqpait (a system of symbols first developed to transcribe the Cree language), to teach myself to read and write Inuktitut.

As the youngest, I also got plenty of attention from my siblings. In many Inuit families there is often a favored child, and no apologies are made for that—it's just accepted. There was never a doubt that Bridget was my mother's favorite, but Bridget didn't always appreciate this. I think Bridget was especially kind to me, more like a doting mother than a sister, because she knew that she was favored over me. And both Bridget and I got lots of affection from our big brothers, along with a strong dose of teasing. As they were closer in age, Charlie in particular teased Bridget throughout our childhood. It was part of showing his love for her. I remember when we used to get Oh Henry! bars from time to time, Bridget and I would savor ours by eating them layer by layer: the chocolate first and then the peanuts, leaving the nougat center for last so we could freeze it in the snow and eat it as a kind of Popsicle. We quickly realized that our brothers had caught on to what we were doing and would attempt to find our treats, so we would sneak out of the house and hide the nougats someplace in the snow where they wouldn't be found. But the risk of losing the candy was more a tease than a real threat. Elijah and Charlie were always kind and generous to both of us, and I cherish my memories of those days with my brothers.

BY THE TIME I WAS BORN, the hardships that my mother and grandmother had gone through in their early lives had been largely overcome. I came into a family that was strong and dignified and that had moved beyond their initial struggles in the rapidly transitioning culture. My uncle Johnny had a good job, working for the HBC store, and his family was always nearby. His children, Willie and Annie, and my own siblings and I all grew up as a close-knit family. My uncle was a strong presence for both families, as he was confident and wise, with a calm demeanour. His wife, my aunt Louisa, was also an important part of my life, showing me much affection, in part because I'd been given the middle name Jessie, her mother's name. In our culture, a namesake carries not only the name but also the spirit of the person. A person named after a family or community member who has passed away is addressed, treated, and loved as though he or she were that very person. I later named my first child after my aunt Louisa.

With my immediate family, as well as my extended family, I grew up in a world of safety. The home that my mother and grandmother created was a place of security, comfort, and peace. In traditional Inuit culture children are never hit, nor are yelling, threats, or harsh punishments used to discipline kids. My mother might have been stricter than some parents in our community, but she still followed the Inuit gentle way with children. Disapproval usually amounted to a stern look or a clear verbal expression of disappointment from her.

But as in most Inuit homes in those days, there were many unspoken rules and expectations. In most cases, we children didn't call elders by their first names, but rather by their relationship to us: aunt, uncle, cousin, and so on. And adults would address us by the relationship our namesake had to *them*. In this way, children might be addressed by adults as "mother" or "grandfather" or "sister." We also understood that we shouldn't be too inquisitive or interrupt adult conversation. We would never have spoken back to my mother or to any of our elders, or even questioned what they said. And we weren't to be loud or boisterous at any time. All the children I knew understood this completely, since as a norm grown-ups weren't constantly telling us to pipe down. Rather, we learned how to behave by watching the adults around us.

Inuit have always placed a high value on the ability to be calm, controlled, focused, and reflective. To be a successful hunter, you can't be noisy. During long hours out on the ice and snow, hunters have to remain quiet so that the animals are not driven away by the sound of human voices. Even physically, we Inuit have learned the importance of quiet: the ability to remain still is an essential survival skill on a hunt. These habits have become part of our social behavior, too. I remember many times our aunts and uncles, friends, and neighbors would come to visit, yet despite our home being full of people, there would be many quiet moments. People did not talk simply to talk. They might read a magazine, look at pictures or look out the window. Silences were accepted, companionable and comfortable. As children, we learned by the example that the adults set.

We children were taught other things by watching and listening. Starting around the age of ten, girls would learn to sew. The women in my community were wonderful seamstresses, and still are. I remain in awe of the younger generation's ability to produce both practical items and decorative pieces of great beauty. During my childhood, however, the women would hand-sew *amautiit* (baby carriers for mothers), parkas, and boots out of sealskin and caribou hides, cleaning and softening the skins by hand. Heavy wool duffle socks and the inner part of the *amauti* and parkas were also sewn by hand. The Grenfell cloth *silapak*, the outer part of the parka or windpants, which was filled with down, was sewn by machine. Traditional clothing was supplemented with "southern" clothing purchased at the Hudson's Bay store. My mother, like so many Inuit women, was always a stylish dresser, and she made sure that her children were well outfitted too. I have a picture of my sister and me, in tartan skirts and crisp white blouses, our hair in pigtails and ribbons. In another family portrait, I'm wearing beautiful handmade sealskin *kamiit* (boots). Bridget has on a sweater and I'm wearing a Mickey Mouse sweatshirt, even though I didn't have a clue who he was. My favorite photo of my aunt Louisa shows the kind of modern flair she shared with my mom and several other women of the community: her hair is swept up in a braid that wraps behind her head; at her throat is a dramatic jeweled brooch.

While young girls learned to prepare the animals for food and for making clothes, young boys would be taught the skills necessary for a successful hunt, including how to build *qamutiik* (sleds) and make *illuvigait* (igloos). The wooden sleds required extreme precision in the cutting of the wood and the tying of the ropes. The snow houses were works of extraordinary ingenuity that showed the impressive engineering skills of our people. Properly made, *illuvigait* can withstand the weight of a polar bear atop them without collapsing and are warm enough to sleep in or for a mother to birth and nurse her baby. Elijah and Charlie got quite good at carving the blocks of snow and assembling them to make the shelters. I recall them building small *illuvigait* as playhouses for other children, including Bridget and me, in the winter.

While girls and boys were taught different things, we all learned to fish from an early age. I was about three when I started going out on the ice of False River with my family to fish for trout. I loved traveling there by *qimutsiit* in the winter. In the summer, many families like ours would pile into canoes to travel the Kuujjuaq River for fish, seals, and whales, as well as to gather eggs from ducks, geese, and seagulls.

The hunt was another activity that united the whole community. By the time I was born, my older brothers had already been taught how to hunt by relatives, including my uncle Johnny, who himself had learned from older relatives. I'd watch my brothers as they got ready to leave for family hunting trips. The precision with which they would prepare the *qamutiik* was fascinating. They first cut damp peat moss from the soil in squares and brought it home, where they checked it for any small stones. My brothers tended to do this the modern way—with their hands—but traditionally hunters would have put chunks of peat moss into their mouths to find and remove the stones. Once the peat moss had been cleaned, they shaped it around the bottom of the sled runners. Then they cautiously dripped small amounts of warm water along the moss. Traditionally, this water would have been warmed in their mouths before being applied. They did all this outside in the freezing cold, the steam rising up in little clouds from the moss as they worked. While the water was forming a layer of ice on the moss, my brothers polished it with a small piece of wet caribou fur until everything was frozen hard. They repeated the process until several layers of ice had built up on the runners, sometimes smoothing the ice further with a plane. Once the peat had a thick, hard surface of ice, the runners would glide easily on the ice and snow. Next, Charlie and Elijah worked on tightening all the pieces of the wooden *qamutiik* by binding them together with strong rope. Finally, they tied all our gear and belongings, as well as the caribou hides we used as seats, to the sled. During this entire process, Charlie and Elijah worked with absolute focus and attention. It was almost mesmerizing watching them labor in the same meticulous way that Inuit had worked for generations.

I learned from watching my brothers that safety was everything. You had to be extremely careful and precise in these preparations, both for your own welfare and for that of your family, who relied upon your ability to lead the dog teams out onto what most people would consider a harsh, difficult, and rough terrain. My brothers had learned well from my uncle Johnny and others, and I always felt safe in their competent hands as they led us out onto the land.

Because we had no father in the family, my brothers grew up quickly, taking on adult responsibilities very young and providing for our family. They hunted for ptarmigan, geese, seals, and whales. When they went hunting for caribou they could be gone for days or even a couple of weeks, sometimes coming back empty-handed. When they went fishing and ptarmigan hunting, they often took my mother, grandmother, Bridget, and me, making the trip a family event. Our favorite fishing grounds were on the Atsaasijuuq, or False River, named for the way the ice slid when it broke apart in the spring. Although the men would fish, too, the women and children remained fishing as the men went off on snowshoes to find ptarmigan in the small trees and willows. I have many fond family memories of that place and those times. Wherever we had been to hunt and fish, the end of the day was always wonderfully exuberant as we joined together along the river, near the trees, to build a fire and enjoy a communal meal, eating what we had harvested.

Even when my brothers or my family hadn't gone on the hunt or caught anything, we shared in the successful hunts of others.

It's hard to describe the excitement that would flash through Kuujjuaq when word came that hunters were returning with a large harvest, like a seal or a whale or a caribou. Word spread from neighbor to neighbor, from house to house, and everyone headed to the home of the hunter. If the men shot a caribou, they would cut it up on the land and then distribute the pieces among the community. If, however, the kill was a seal, people would gather in one place, usually the hunter's house. The women and some of the men would spread pieces of cardboard on the kitchen floor and lay the animal on top of it. Standing and leaning over the animal, usually the hunter would cut open the seal along its belly, exposing first the nice layer of fat and then the meat. Next, the entire carcass would be opened, exposing the delicious organs. Sitting or squatting on the floor, the men and women would begin to cut up the carcass with sharp knives or an *ulu*, the Inuk woman's traditional multi-purpose knife. (The *ulu* can be used to cut meat or scrape the fatty layer off the sealskin, and to eat with. Today we still use the *ulu* for the same purposes, as well as for chopping vegetables.) Everyone else, including the children, would sit circling the seal. Pieces of meat would be passed around for the children and other adults to eat, women often getting the parts that were considered delicacies. The liver was one of my favorites. But the best moment was when we would reach into the open cavity of the butchered seal and dip our hands, coating our scooped fingers with sweet, rich blood, which we licked off like honey. It was butchering and a feast all at once. Those precious moments, sitting on the floor with my grandmother and mother, my brothers and sister, my uncle and his family, and so many members of my community to enjoy our traditional foods were treasured times.

From an early age, I loved my country food, which is what we call the food that the land around us provides: seal, whale, caribou, goose, duck, ptarmigan and, of course, fish. Along with *muttaq* (whale skin with a layer of blubber below), my favorite food was and still is *quarq*, which is frozen caribou or fish dipped in fermented seal oil (*misiraq*). It's like dipping lobster in drawn butter. (That said, when the salmon were running, we would have salmon every day for lunch and dinner, and I got pretty sick of it.) We supplemented the protein of our country food with kelp and, in the summer, fresh berries. I still crave and eat country food whenever possible. And to this day my brother Charlie sends me caribou, or *tuktuviniq*, from Nunavik, and my godson Jon and other family members send me goose and fish.

But the importance of country food to my community goes far beyond taste and individual preferences. Country food is the fuel we need to thrive in the Arctic. Full of omega-3, fat, vitamins, and minerals, foods like seal meat actually keep a person warm—even if the meat is frozen. In fact, eating much of this food frozen or raw is one of the ways we get the maximum benefits from it. *Muttaq*, seal liver, and brain with the fat, for example, are full of vitamin C, which is destroyed if these foods are cooked. Many of the country food staples, including seal, are so nutrient rich that they are pretty much a whole food, providing everything we need to stay healthy.

When I was young, even if there wasn't a bountiful hunt, we always had enough to eat. My mother was a great baker and cook, and she often prepared meals for the nursing stations and mining companies. While ptarmigan and goose were usually available, when we didn't have fresh fish or meat, my mother and grandmother would prepare southern food with groceries purchased from the HBC store: pork chops; macaroni with canned tomatoes; Klik luncheon meat; canned chicken, bacon, and corned beef; canned or powdered milk. Dessert was often canned fruit and evaporated milk. While our southern food choices were somewhat limited, from time to time I'd be introduced to something new.

On one occasion, when I was about three years old, the Hudson's Bay Company store in Old Fort Chimo almost burned to the ground. The whole community tried to save it because it was so important to us. Although very young, I could sense that this was a real trauma for the community as I sat perched on the windowsill of our house, watching as the adults frantically tried to remove supplies from the store before losing them. After the fire had been put out and everybody was still milling around outside, my mother took me to the site. We stood together, watching the smoke as it continued to billow out of the building.

My young cousin Mark was there as well, with his mother, Christina. She was my *sanajik*, meaning the one who had assisted at my birth and cut my umbilical cord. Christina reached into a bag and handed her son a little round burned-looking thing. He put it in his mouth and seemed to be enjoying eating it so much that I wanted one of the burned things, too. Christina handed me one, and I popped it in my mouth. In those early years, it was rare to get any kind of fruit or vegetable that didn't come from a can, so I assumed the blackened fruit I was eating had come from a can in the fire. It would be years before I discovered that it was a dried plum.

OUR FAMILY HUNTING AND FISHING TRIPS were an important part of my early childhood, but they represented only a fraction of the time I spent outdoors. For the children of Kuujjuaq, the great northern landscape was our playground. We explored the natural world, picking berries, looking for birds' nests, and playing on the hillsides surrounding our town in the summer. In the winter, we would sometimes slide down those same hills on sealskins. My friend Martha Koneak's mother, Joanna, would ask us to head to the hills with her sealskins so the snow would make the skin shiny and pliable before she sewed *kamiit* with them.

When we could, we played outdoors from morning until night. As Fort Chimo grew, parents became a little more concerned about letting children play outside after dark. If we stayed out too late in the evening, my mother and other parents, like my friend Lizzie's father, Johnny Saunders, would walk through the streets with flashlights, looking for their own children and scooting everyone else back to their homes. For a number of years, a horn also blared at 10 P.M., signaling that it was time for young people to be indoors. I believe this alarm system was the work of the federal day school in Fort Chimo and the Department of Indian and Northern Affairs. It did get us moving inside, but it was the attention of the adults in the community that made us feel safe and secure.

I'm still struck by how important this sense of a supportive community was to me in my early childhood. My family may have been fatherless, but we were never made to feel ashamed— nor *did* we feel any shame or lack in any way (although my mother and her siblings may have felt differently in their childhoods). The only time I recall feeling uncomfortable about being raised by a single mom was when a little girl from one of the French-Canadian families living in Kuujjuaq told me she couldn't play with me because I didn't have a father. And while I've heard from others that some from the older generation were treated badly because they were marked as *qallunaangajuk*, meaning part white, I don't recall ever having this experience. I was always addressed with terms of endearment, and I often heard those wonderful little murmurs that my grandmother made from the other adults around me.

When I was six, I joined my older siblings at the newly built school in New Fort Chimo. The school was bigger than most of the buildings in town, but for my first two years, I was in the same classroom as my siblings. Eventually, the students were divided into smaller classrooms by age and grade.

By the time I started school, Charlie and Bridget were proficient in English. And, of course, my mother spoke good English. But my grandmother spoke only Inuktitut, and we spoke nothing but our mother tongue at home, so I knew no English at all. Like many in my class, however, I learned to speak, read, and write in English quite quickly.

I recall clearly that every morning at school, we would be served soup or hot chocolate and crackers along with vitamins. Oh, we tried so hard to swallow those vitamins without biting them, they tasted so horrible.

Some teachers were quite strict and gave kids who spoke out of turn or fooled around in class the strap or a ruler on their hands. Sometimes they put children in the corner.

But physical punishments were not the only exposure we had to a new southern world view at school. Our first readers were books from the "Dick and Jane" series, which were completely disconnected from any reality we knew in the Arctic. Dick and Jane's dog, Spot, for example, bore no resemblance to the working dogs we had grown up with. The suburban houses with picket fences and leafy trees looked very foreign to us indeed.

Other southern things we were introduced to, however, were much more enjoyable. We learned games like dodge ball and "Andy Andy Over," which involved throwing a ball over a building. And we had such fun at the end-of-the-year field day. Most Inuit children are natural athletes, and many of us, including me, thrived in those sports activities. I lived for the races and games at the end of the year when we would get red, blue, or yellow ribbons for our efforts. My friend Martha, with whom I would go sliding, and I excelled in those sports together.

The teachers, for the most part, were very good. And while they were all anglophone qallunaat, they and their families were very much a part of the community. They would invite us to their homes, where they showed us how to cook southern food and bake cookies. They would teach us songs as they played the piano and organize community socials and music nights, complete with food like baked beans and bread.

Yet while we enjoyed learning about these southern pastimes, the transition into the modern world was hitting us quickly. In the past, the news of the world around us had drifted in only by bits and pieces. Sometimes this was really confusing. I remember when John F. Kennedy was assassinated, the elders and adults of the community spoke about how he had been a president who kept wars from happening. We children did not understand what that meant, and automatically assumed that war would break out the next day. We fearfully looked up to the Arctic sky, waiting for the airplanes, the fighter jets, to arrive and for war to start.

But as we grew older, and as Kuujjuaq grew, the outside world became more and more a part of our world. There were more cars, more institutions, more *qallunaat*, more nine-to-five jobs. There was less hunting (largely because of the nine-to-five jobs), less freedom, and less independence. And while Kuujjuaq and the North were changing, for me and a number of other young students, the transition would become even more abrupt.

After I finished my fourth year at school, my mother took me aside to tell me that my friend Lizzie and I had been chosen to go south to attend school. I didn't realize it at the time, but we were to be part of a federal government program that selected "promising" Inuit children with a potential for leadership to be educated outside of the Arctic. Lizzie and I, my mother told me, would be joining the Rosses, a nursing couple my mother had worked for, in their home in Blanche, Nova Scotia. Although I really had no idea what this all meant, I was excited at the prospect of a new adventure. It would, however, turn out to be the end of my Arctic childhood of ice and snow. And it would mark the beginning of a series of losses that I would struggle with into adulthood.

2

FROM DOG TEAMS TO MINISKIRTS AND ROCK 'N' ROLL

MY DEPARTURE FOR NOVA SCOTIA was not the first time I had left home. I'd made my first trip south when I was eight or nine. I had recurring tonsillitis, and the clinic in Chimo couldn't perform the surgery I required. To this day, those of us in remote communities without hospitals have to travel great distances to get medical care, and that often means going by plane. Along with several others from Kuujjuaq, I was sent by St. Felicien Air Services to Roberval, Quebec, to have my tonsils removed.

I was to spend an entire month away—two weeks in the hospital following the surgery, and two weeks recuperating and waiting for the plane to bring us back home. Regular flights to the North weren't scheduled in those days, so we had to wait until the plane had a reason to come back to Roberval, either with other patients or with paying passengers. Joining me on the trip were three little boys from my community, who were all younger than I was. Two of the boys, Johnny Adams and Joseph Saunders, were about four or five, and Matthew Sequaluk was still a toddler, not yet walking. Although two women from the community had also come down with us for medical treatment, once we got to the hospital they were on another ward, and we rarely saw them during the first two weeks. At the hospital, Johnny and Joseph were longing for their mothers and cried all the time. My maternal instincts have always been strong, and even at that young age, I felt protective of those little boys.

Although I, too, was homesick in this foreign place where I didn't understand the French language, I tried hard to be brave. I would wait until the little ones were napping, then go sit in a rocking chair in front of the window and cry silently. One day, while I was in that weepy state, I heard a little sniffle beside me. It was Johnny. When he saw me crying, he asked, *"Qiagiviit?"* ("Are you crying too?") We both cried together. Since that day, Johnny and I have remained close friends, and we have an unspoken pact to always be there for each other. Indeed, in later years, Johnny was to be there for me during difficult times.

After our release from the hospital, all six of us, the four children and two women, were brought to Pointe-Bleue, a little First Nations community north of Roberval, where we all lived for the next two weeks with a First Nations woman named Mrs. Robertson while waiting for our flight. I felt more at home there but ate barely anything, as my throat was sore from the surgery and I didn't like the food. I was very fond of my country food, but there had been none of this familiar fare for the many weeks I'd been away. The absence of this comfort was especially hard since I was still struggling with homesickness and missed my mother and grandmother terribly. One day, however, a relative of Mrs. Robertson's brought some moose meat, which she fried up for me. It reminded me of caribou from home and was the best thing I'd eaten in a long time. I finished my entire plate, even asked for more, and started to feel better all around.

During our stay in Pointe-Bleue, we children found the presence of the two older Inuit women, Lizzie Gordon and Mary Tukkiapik, comforting, but the trip was traumatic just the same. (Mary died not long after returning home, I suspect from a terminal illness, but Lizzie lived for a long time afterward and passed away only recently.) To undergo painful medical procedures while severed from our families and surrounded by people who didn't speak our language filled us with fear, sadness, and longing. We were too young to appreciate that we were there for our health and future well-being. That month felt like a year to me, and I remember counting the days we'd been away and wondering why the plane didn't come to take us home.

Finally the flight home was arranged. When we landed in Fort Chimo, all the mothers gathered at the airstrip to meet us. We were so excited to see familiar faces and hear everyone speaking our language again. I recall talking with Matthew's mom, Sarah. Matthew had learned to walk while we were away. I told her that I had taught him how. It occurs to me now that as much as we children suffered while away, our families must have worried and longed to be at our sides as well, especially the mothers of the young children.

MY SECOND EXPERIENCE GOING SOUTH was much more pleasant. My mother, being an interpreter for the nurses, was asked to travel on the *C.D. Howe* medical ship in the summer of 1962. The *C.D. Howe* traveled up to the small communities of the High Arctic to test people for tuberculosis. Inuit patients who were diagnosed with TB were then brought aboard the ship and taken to southern hospitals. My mother was to be on the ship for the entire summer, so my sister and I were sent to Montreal to spend that time with Hal and Myrtle Tincombe, a couple whom my mother had befriended through her work at the nursing stations. This generous and kind-hearted couple took in Inuit patients who were sent to Montreal for treatment and looked after them as they waited for appointments.

The Tincombes lived in Lachine, Quebec, a suburb of Montreal, but they also had cottages in Vermont and Ontario, so our summer days were divided between those places. Despite the motion sickness I suffered whenever we had to take a long car ride, I loved our trips to the cottages. I learned to swim, enjoying the warm waters and the absence of all the mosquitoes we had up North. I was a cautious child who didn't like taking big risks, so diving under logs and all the little adventures we had in our new surroundings built my confidence and helped me to become a bit bolder. I also made new friends in Lachine and at the cottages. I was a year older than I had been when I went away for my surgery and was getting more familiar with "the South," acquiring a taste for southern food and treats. That summer away, then, was good training for the years to come, when I'd be apart from family and community for much longer periods of time. And while I would never get used to being away from my mother, grandmother, and siblings, during that summer with the Tincombes I had Bridget with me, which was a huge comfort. What's more, during those months I came to love Mr. Tincombe, with his humor and gentleness. He was the only man I have ever called "Dad" in my life. (I was tearfully able to tell him this many years later when I visited him in the hospital, his daughters Carolyn and Diane by his side, after he'd suffered a stroke.)

As soon as the summer was over, I went home for another year of school in New Fort Chimo, but Bridget headed instead for school in Victoria, British Columbia, with an Anglican missionary couple. My mother, after her summer on the *C.D. Howe*, accompanied my sister on the journey by train from Montreal. My mother wanted her children to have a good education. I assume she met the Anglican minister and his wife through the Tincombes, and talked with all of them about having both her daughters educated in the South. I recall my sister crying as she was leaving. Now that I think back to that time, I realize how difficult that must have been, going so far away with people she didn't know.

The following year, I was told that I would be heading to the opposite end of the country. I'm not sure how the plan came about to send Inuit children who showed potential for leadership to be educated in the South, nor how I came to be a part of it. Many years later, a number of us who'd been sent south from Nunavik to live with *qallunaat* families realized that we must have been part of a larger government plan. Thanks to Zebedee Nungak, Eric Tagoona, and Peter Ittinuar, who shared their experiences in a documentary called *Experimental Eskimos*, we eventually would understand that the government of the day believed this southern education was an important step in training young Inuit to be future leaders of their communities.

Lizzie and I were told that we'd be going to live with Joseph and Peggy Ross in the small town of Blanche. Prior to moving back to Nova Scotia the year before, the Rosses had been living and working in the communities of Nunavik, particularly on the Ungava coast, for about eight years. During their time in the Arctic, my mother had worked with them daily, as an interpreter as well as a cook. I had come to know them and their son, George, who was close to my age. Perhaps because of my happy summer with the Tincombes, perhaps because of my familiarity with the Rosses, I wasn't at all worried about joining this *qallunaat* family and going to school so far from home. I was only ten, after all, and naive about what it really meant to be away from my mother, grandmother, and community. I was in for a brutal shock.

The journey south was traumatic, foretelling of the struggles soon to come. The aircraft we traveled on in those days, either DC-4s or Super Constellation propeller planes, were not like the fast-moving jets of today. The flight from Fort Chimo to Montreal, our point of contact to the South, took seven to eight hours. I quickly became extremely motion sick on the long flight—so ill, in fact, that I had to put on an oxygen mask. In Montreal we boarded another flight to Halifax. That was followed by a four-hour drive to Shelburne County and Blanche.

But my severe motion sickness throughout the journey was nothing compared to the heartache of being separated from my mother and grandmother. The stark reality of being alone with Lizzie struck me suddenly on the plane to Montreal. It hit me that, in heading south, we were leaving our family, our culture, and our community—and for more than a few months. Here we were, barely ten years old, disconnected from family and community. But as upset as we were, we had no idea that this sudden severing of culture would end up marking us for a good part of our lives.

After arriving in Blanche, I was bedridden for the next three days, unable to eat. I eventually got over the nausea, but my discomfort continued. The Rosses felt that the more we adapted to southern ways, the better it would be for us. Our diet, it seemed, was an important part of this adjustment. Of course, there was no country food available. This proved to be exceptionally hard for me. Of all my siblings, I had spent the most time with my grandmother and, therefore, had become the most attached to the kinds of food that she prepared and ate: *muttaq*, seal stew, frozen caribou, and fish with fermented seal oil. For me, country food was a powerful part of feeling at home. At the Rosses', not only was there none of this comfort food but even the southern food that was served was different from what my mother and grandmother prepared. In particular, I hated the fresh peas and the fresh milk. If I ate peas at home they were from a can, and milk, which came tinned or powdered, was used only on cereal or canned fruit and added to tea. But in Blanche, the Rosses insisted that Lizzie and I drink tumblers full of the stuff, which I found revolting.

Lizzie and I also were made to dress a specific way. I remember wanting to wear a sweater with a dress and being told that was not how one dressed there. We had to be prim and proper.

The Blanche community was also both isolated and isolating. There were only five families in the tiny coastal town, and only one child other than George Ross—a boy named Frankie who lived with his single mother. We were all bused to another small community, Port Clyde, to go to school, and while Lizzie and I did make a few friends there, we lived too far apart to see them after school. So although we spent some time outdoors and walked in the woods behind the Rosses' house on occasion, we didn't fill hour after hour playing in the open air or roaming the hills to find birds' nests and pick berries, as we had in Fort Chimo. Instead, we did chores around the house or on the Rosses' small farm, or we did homework.

Even more difficult to adjust to than our new diet and routines was the family dynamic in the Ross household. In the Inuit homes that Lizzie and I had grown up in, children were expected to be well behaved, relatively quiet, and mostly self-reliant. And certainly my mother was a strict, commanding presence. But Mr. Ross's treatment of his son, George, was more than strict. Lizzie and I felt the tension between father and son, and were unnerved by Mr. Ross's displays of anger. We could also see how difficult it was for George to please his dad. We felt bad for George, and far away from our Inuit childhoods.

The enormous gulf between our home and the new world we found ourselves in was made worse by our inability to easily talk with anyone in Fort Chimo. The only way that anybody could communicate with our community from outside the region was by ham radio. To make a call to my family, I had to connect, via phone, to an operator working the radio in a Fort Chimo office building. The ham radio set-up was like talking on walkie-talkies. I would have to say, "Hello, how are you? Over." Then my mother would speak. We were never able to talk at the same time. This awkwardness was made worse by Blanche's rudimentary local telephone system, where residents shared a common line. This meant you could use the phone only when no one else was on the party line. It was therefore impossible to just pick up the phone and call my mother to tell her how desperately I wanted to come home or that I was not well. And there was certainly no privacy on either end when we did talk.

Lizzie and I were desperately homesick. The Rosses attended to all our physical and educational needs, but they didn't provide the love and affection, the family warmth and sense of community that we wanted and needed. We cried for two weeks non-stop. And we wrote anguished letters home, telling our families that we wanted to return.

Lizzie would send letters to her sister, Annie, most of the pages filled with, "I want to go home. I want to go home. I want to go home." I'd write letters in English to my mother and other family members, including my sister-in-law, Ida, who had married my brother Charlie. As my grandmother didn't read English, I would write to her in syllabics, in Inuktitut. We would put all these letters in an envelope and send them off. At least, we thought they were being sent off.

One day, after we had written letters and asked Mr. and Mrs. Ross to mail them for us, we came home from school to find these same letters opened and spread out on the dining room table. They had obviously been read. Even though I was very young, I felt my privacy had been grossly violated. But the experience got even worse. The Rosses told us that we were never again to write letters home unless they had screened them first. In essence, we could not communicate freely how we felt to our families. The only letters that were never screened were those between my grandmother and me, as they were written in syllabics. I cherished my grandmother's letters in response, and I still have them today, stored in my trunk for safekeeping.

To be fair, despite the harshness of their actions, I believe the Rosses' intentions were honorable. They certainly never set out to be harmful or hurtful to us. In fact, we called them Uncle Joe and Aunt Peggy. But they felt that they needed to educate us and push us to excel in school, which they assumed depended on assimilating us into southern culture. And the Rosses also clearly felt it was their job to ensure that our families didn't worry about us.

But the effects of this censorship were profound. It took me a long, long time after that experience to feel comfortable or even *able* to express myself, my thoughts, and my feelings. In one simple act, the Rosses helped to weaken my voice for years to come.

However, the Rosses did try to expose us to southern experiences they thought we might enjoy. When Lizzie and I snared a rabbit, Mrs. Ross used the meat to make us rabbit pizzas, of all things. And while I had seen TV at the Tincombes' (TV wouldn't reach Fort Chimo until the early eighties), at the Rosses' we were allowed to regularly watch shows like *Gunsmoke* or *Don Messer's Jubilee*. The Rosses even took us on a trip to Boston via the ferry *Bluenose*. While there, we went to see the filming of a children's TV show, which I enjoyed—that is, until the children in the audience were instructed to get up and perform something called the "Freddie dance." Being an introvert at heart and shy to be acting silly, I was reluctant to join this public display and stubbornly remained seated. Lizzie kept me company on the bench. Mr. Ross wasn't pleased at our lack of participation, and the ride home from the television studio came with a lecture about being obedient and joining in the fun. It wasn't even close to being fun for me.

Although we went home for a couple of months during the summer, we had to spend Christmases in Nova Scotia. That first Christmas was strange. While the Rosses had similar Christmas traditions to ours and there were plenty of presents under the tree for Lizzie and me, I missed waking up to the sound of my grandmother's radio playing in the kitchen. And I longed for the communal warmth of our Kuujjuaq festivities—the steady stream of neighbors through the house, the handshakes and party games, my mother playing the accordion at the community square dance. But by Christmas time, Lizzie and I had both come to realize that expressing our homesickness or our unhappiness would serve no purpose. And while we didn't consciously numb ourselves, we let the days pass in a blur, moving from school to chores and homework and back again.

Indeed, under the Rosses' strict guidance, we did excel at school, quickly becoming proficient in our second language. In fact, we often outperformed our English-speaking classmates. To this day, I still have the silver dollar that I received for being at the top of the class.

But our achievements at school came at the cost of our Inuit knowledge and skills. Lizzie and I did not speak Inuktitut all the months that we spent with the Rosses, and during those two years, we lost a remarkable amount of our mother tongue. I remember trying to practice Inuktitut before heading home for the summers, so I wouldn't feel so lost with my family and community. But it proved to be surprisingly challenging, and sometimes embarrassing. Many people back home couldn't understand how we could lose our language so quickly. However, at such a young age, spending an entire year not hearing any Inuktitut made it extremely hard to hold on to it. In fact, it would take both Lizzie and me most of our adult lives to regain our mother tongue. But that was not the only cost to our "Inuitness." During the years we would spend away from home, other girls our age were learning how to prepare country food, embroider, and sew. They were mastering the cultural skills of working with caribou hide, sealskin, and goose down. When Lizzie and I would return to the Arctic, we'd see family members and friends who were being taught how to create beautiful, practical Inuit clothing and realize how much we were missing. (To this day, Lizzie and I are not passionate about sewing, unlike many of our fellow Inuit women our age. We still pick berries together, eat our country food with gumption, and are attentive and caring mothers and grandmothers, but we don't sew.)

While both Lizzie and I found losing touch with our Inuit lives extremely difficult, I was faced with another challenge to my Inuk identity. My time in Nova Scotia made me more aware of the fact that I didn't *look* Inuit. When first meeting Lizzie and me, neighbors and friends of the Rosses would often express surprise. "I thought you were getting two Eskimo girls," they might say. "But *she's* not Eskimo," they would add, looking at me. It was a shock. My world was the Arctic, my people were Inuit, I was Inuk. But suddenly others were questioning my identity, challenging my sense of self. Not only had I been uprooted from my home but now people seemed to be saying that it had never been my home in the first place. I found it profoundly upsetting and began to dislike my *qallunaaq* look.

The comments about my appearance also made me insecure in another way. Some of the kids at school would say things like, "Are you really Eskimo? You look white." That would lead to questions about my parents. It was too difficult to tell them that I didn't know who my father was, that he hadn't stayed to marry my mother. Instead I'd say, "Oh, he's dead." It was easier than explaining my situation, but unfortunately this response only seemed to reinforce the idea that there was a lack in me.

During the second year in Blanche, I also experienced one of the worst losses of my life, and yet it was one that I couldn't grieve until I was in my thirties. Indeed, when it happened, it was probably hard to tell that anything momentous had occurred. The phone rang, and Aunt Peggy went to answer it. When she came back into the living room, she asked me to come into her bedroom with her. There she told me that my grandmother had just died. I don't even remember saying anything in reaction to this news. We left the bedroom, and that was it.

I knew my grandmother had been diagnosed with an illness, a cancer, which I would not have fully understood at that age. She had, apparently, been sent to Roberval, but the cancer was too far advanced for treatment, so she was sent home. I suspect she knew that she wouldn't see me again after my first summer at home because she was quite concerned about when I'd be leaving for Nova Scotia.

My grandmother had always been so much more than a grandmother to me. She was my other parent. And she was my foundation. My grandmother gave me a command of my culture, my language, my food, my Inuk identity. I had spent all my early years by her side. She raised me. And yet, I don't remember crying at the news of her death. There were no loving and understanding arms to fall into for support, and I couldn't take part in the community grieving that would have helped to open up my heart when we buried my precious grandmother Jeannie. It was as if being sent away had shut down my emotional responses, as if the acceptance that I'd been forced to embrace had muted everything for me. Life just went on.

While I was unable to grieve the loss of my grandmother, even after returning to Fort Chimo the following summer, I wasn't entirely silent about my unhappiness in Blanche. At some point during those months at home, I shared with my newly acquired sister-in-law, Ida, whom I'd become very close to, the discomfort I felt about going back to the Rosses'.

My sister-in-law talked to Charlie, who, as a true big brother, had always been protective of me. He, in turn, told my mother that he didn't think they should send Lizzie and me back to Nova Scotia. My mother spoke with Lizzie's parents, Johnny and Elsie Saunders, and they went to talk to the Department of Indian Affairs agent together. After talking with our parents, the agent asked to speak with Lizzie and me about why we didn't want to go back to Nova Scotia. We shared, as well as young children could, the tension that existed in the

Ross household, while trying to avoid portraying the Rosses as bad or unkind. I don't recall the details of that conversation, but I do remember the feelings I had about having to deal with all that at such a young age. Although the agent wasn't aggressive or threatening, in those days government officials had a lot of power, so it was hard not to feel frightened or distressed. Being asked questions by this man of authority made me feel as if we were being grilled, that we were the ones in the wrong. Yet, I think the official was trying to be compassionate. I recall him saying that we were not in trouble and he would report that we wanted to be with our oldest sisters, who were on the list to be sent to the Churchill Vocational Centre.

In the end, Indian Affairs decided that the following year, Lizzie and I would be sent to Churchill, Manitoba, where the government had established a residential-type school for students aged twelve to seventeen. At twelve, I was to be the youngest student there that year.

As we were preparing to leave for Churchill, a pivotal moment in my life occurred. Jamie Clark had been the Anglican minister in Fort Chimo since my early childhood. He knew me, my mother, and my family well—in fact, he'd baptized me, and I called him Uncle Jamie. He was the official who prepared the identification papers that we were to bring with us to Churchill. On my card, Uncle Jamie had written my name, my date of birth, my mother's name, *and* my father's name—George Kornelson. At the age of twelve, it was the first time I had seen or heard my father's name. I had never thought much about my father during my younger years. There were the odd times I would attempt to ask my mother about him, but her responses were always brief and not very enlightening, so I learned not to inquire about him. Most of the time my fatherlessness seemed beside the point. I was an

Inuk child, being raised in an Inuk household. It was only after the French-Canadian children refused to play with me that any seeds of inadequacy started to take root.

Uncle Jamie told me what I had already known, that my father had been a Royal Canadian Mounted Police (RCMP) officer working in Fort Chimo. Uncle Jamie and my father had, in fact, grown close working together, performing a wide range of services in the Arctic that would otherwise have been absent. My father and Uncle Jamie would travel up the coast into the smaller communities, Jamie working as a missionary, and my father bringing syringes to give vaccinations or penicillin shots. In this way, they wore a number of hats beyond the typical roles of their professions. Because of their friendship, Uncle Jamie had known who my father was since I was born. And I think because of that, he had always treated me with love and a sense of responsibility for my well-being.

Hearing this news from Uncle Jamie, a man who cared for me and whom I trusted, was incredibly powerful. I knew it had to be true. I remember walking home with an upbeat step that day. I had been given the name of my father. It was the best gift I could have received.

And now, to have this new information as I headed to Churchill with my sister, Lizzie, and her sister, and many other fellow Inuit teenagers from Northwest Territories (part of it now Nunavut) and northern Quebec, was incredibly affirming.

THE PLANE RIDE TO CHURCHILL wasn't traumatic like the one I had taken at ten when leaving home for Nova Scotia. Our flight would be on a chartered aircraft. Lizzie and I were together again, but this time we had the familial comfort of our sisters and the company of many other Inuit teenagers. And Lizzie and

I had already heard so much about the Churchill experience from our sisters, who had been there the year before. While there was excitement in the air as we boarded the flight late at night, the flight itself was uneventful—I suspect we slept through most of it.

We landed in Churchill in the middle of the night, tired from the long trip. We were greeted by supervisors, the residential "parents" who would look after us in the dormitories in shifts. They told us to line up immediately. They gave each of us a number and a set of clothing—underwear, T-shirts, shirts, pants, socks, and so on. I was number eight. Then they instructed us to get ready to hop into the shower. Before bathing, we had to undress nearly completely and have lice medicine rubbed into our hair. After leaving it in for the requisite amount of time, off we went into the showers to wash it off. Then on with our numbered underwear and pyjamas and into the dormitory rooms.

We were assigned four to a room. These bare-bones spaces contained two sets of bunk beds and a built-in desk up against the window, where we could do our homework. Lizzie and I took the upper bunks, and our sisters took the lower beds. Our sisters had warned us that a night watchman would come by and check each room while we slept to ensure everyone was in their beds after a certain time. I imagine our sisters didn't want us to be frightened by the flashlight when it shone on us during the rounds.

Remember, we were young children and teens, arriving in the middle of the night in a strange setting. The greeting party were strangers who numbered our clothes, treated us as if we were infested with lice, and lined us up half-naked for showers. There was no dignity in this introduction to our new home, not to mention a blatant absence of welcoming, loving energy. And while the presence of my sister and the other Inuit children made me feel a whole lot safer than I might have otherwise, my introduction to this new institution was still an unsettling one.

The Churchill Vocational Centre opened in the early 1960s and remained so for more than a decade, closing its doors in 1973. During my years there, the center housed about two hundred Inuit students from Northwest Territories and northern Quebec. Although in Canada most residential schools were funded by the Canadian government but run by the Christian churches, this one was operated by a federal department of vocational training in Ottawa. At the time I arrived, the superintendent was Mr. Ralph Ritcey, a tall man who often smoked a cigar. He was based in Ottawa, so we never saw him, but when I did get to know him in Ottawa a few years later, I found him to be a kind, gentle, well-meaning man.

The students called the Churchill Vocational Centre the "hostel," although I'm not sure why. Our dormitories had previously been army barracks.

My sister and the majority of the students went to the actual Churchill Vocational Centre, located in the same building as our dormitories. The federal government had tailor-made the three-year vocational program for Inuit students from northern Quebec and Northwest Territories. The program taught academics as well as vocational skills like cooking, child rearing, and sewing for the girls, and trades such as carpentry, mechanics, and welding for the boys. And while the skills (especially those taught to the girls) were similar to the types of things Inuit children would have learned with their families and in their communities, the instruction was in southern ways. We were being deprogrammed from our Inuit culture and reprogrammed for the southern world.

A few of the students, however, like me, were put into a wholly academic program. When I arrived, I was going into grade six and attended the Hearne Hall School. I went on to do grades seven and eight in the Duke of Edinburgh School, which was considered high school at the time in Manitoba.

Each day, we would get up early in the morning and make our beds, perfectly tucked with corners neat enough to satisfy an army drill sergeant. We then lined up in pairs with monitors beside us, usually one or two senior students and a supervisor, who walked us down the long hallway to the cafeteria, all in silence. There, we lined up for our food, and when we finished eating, took turns sorting the garbage, cutlery, plates, and cups onto separate trays. Then, when instructed by the supervisor, we all rose, queuing to put the trays in the washing-up area, before heading back to our dormitories to get ready for school. We came back to the residence and had lunch and supper in the same regimented manner.

At the end of the school day, we had free time for an hour or so, before repeating our meal ritual. If we did all our homework, we had a little more spare time in the evening, as well. Our extracurricular activities also kept us busy. We loved basketball, volleyball, and gymnastics. Like many Inuit, I excelled at basketball and volleyball, and lived to play sports, which helped to build my confidence. Under the coaching of our gym teacher, Mr. Hamel, I became top scorer for my team. Mr. Hamel then made me captain of the junior team, helping me develop my leadership skills. In grade eight he put me on both teams, the junior and senior, which meant I was able to travel by train to places like Thompson and Flin Flon, Manitoba, to play for the Duke of Edinburgh High School team with the many friends I'd made.

Despite the alienating welcome in Churchill, the three

years I spent there were the best of my teenage life. Along with sports, music and dancing became my favorite pastimes, and being with fellow Inuit kids, many of whom I grew up with in Fort Chimo, made all the difference.

By twelve, I had already learned to love dance, to move to rock 'n' roll. Beatlemania had hit the North, and certainly in Churchill, inside and outside the hostel, it was all around us. Recently I read what William Tagoona wrote on his Facebook page about his Churchill days and the importance of music. His words rang true for me as well: "We embraced this change, we were ready for it." Despite the struggles that we Inuit have faced in our communities, in the past and today, we are a people who have adapted quickly in many areas. In the sixties and early seventies, this new sound was rapidly yet almost seamlessly adopted in the North. We went from dog teams to rock 'n' roll and miniskirts almost overnight. Indeed, our own rock bands proliferated within a few short years of Beatlemania hitting the region. In a sense, for those of us living far from home, the music became the soundtrack of our lives, its words of love substituting for those we missed from family and friends. At the Churchill Vocational Centre, we had our own band we called the Harpoons, with William Tagoona as the lead singer. (William went on to become a well-known CBC journalist in our region, as well as a great musician with several records of his own. He was my first crush as a twelve-year-old. Imagine a fellow Inuk rock star. Who could resist?)

The music of the Beatles and other rock bands became a therapeutic outlet for us in our transition to the residential-school setting. Friday nights were some of the best of our lives because that was the dance and band night. I started to dance not only for my own enjoyment but also as a go-go dancer for some of the student bands—whether they were the Harpoons, with all Inuit players and singers, or others like the Moss or the Burgundy Boys from the community of Churchill. I danced alongside either Susan Tagoona or Nancy Tupiq, my fellow Inuit students, for the Harpoons. For the local Churchill bands, I danced with Rene Camphaug, whom I had befriended along with other *qallunaat* kids at Hearne Hall and the Duke of Edinburgh. Rene's mother sewed all our outfits, including glittery miniskirts that we wore with white go-go boots— short and tall ones.

The Friday night dances were held in the same building as the hostel we lived in. They were exclusively for us Inuit students and the supervisors, although guests were allowed from time to time. Everyone got dressed up for the night. Our families back home helped us buy or order the hip new styles from the Sears and Eaton's catalogs. My mother would send us money or have our catalog orders sent to us at the hostel. How exciting it was when these "in" clothes arrived in the mail! Sporting our miniskirts or psychedelic bell-bottom pants, we did the shake to the Surfaris' "Wipe Out" or one of the popular line dances to the Rolling Stones' "Satisfaction." We did the pony, the monkey, and the jive to the Beatles, the Dave Clark Five, or the Monkees. Friday nights would fly by to the music of our own or local bands, who played all these current songs, or to the hit records of the day.

The dances allowed us to be just like any other teenagers. For those few hours, time stood still, allowing us to forget that we were in an institutional setting far away from home, and that we wouldn't be walking home to our parents after the dance. And it was through rock 'n' roll that we became equals not only with the supervisors who were on the floor dancing side by side with us but also with those around the world who were part of Beatlemania.

Our own communities back home were equally caught up in the rock 'n' roll era, and dances and parties became social events. One of my close friends, Okalik Eegeesiak (who has headed various Inuit organizations, including the international Inuit Circumpolar Council), once told me of the night she took her mother, Simigaq, to a Rod Stewart concert in Ottawa in the nineties. Her parents had been soulmates, partners in life but also on the dance floor. They had embraced the world of dance to rock 'n' roll, as many did during that period of transition. Her mother later became a young widow with many children. Years later, when she decided to take her mother to the Rod Stewart concert, Okalik made sure they got good seats close to the stage. As Rod Stewart began to sing songs familiar to her mother, the tiny Inuk woman stood up and danced, letting out screams of both sadness and joy as she wept and rode a roller coaster of emotions, undoubtedly missing her husband. Hearing that story brought a tear to my eye. I felt that those memories of tumultuous change working their way through Simigaq in that moment could well have been the expression of our collective Inuit world.

While the music and dances were fun, the food served at the hostel was another matter. It was cafeteria fare, and though we got used to it, it's probably not surprising that we developed a taste for junk food to break the tedium of bland meat and potatoes with soggy vegetables. A couple of nights a week, we were allowed to place an order for chips, chocolate bars, and soft drinks or milkshakes using the $1.50 allowance we received every two weeks. Two students would be given the task of taking the order and walking down the long hallway past the cafeteria to the public area, where they purchased the treats from the commissary. (At all times we required a permission slip to go past the cafeteria to this public domain.) We relished those evenings. The other time we'd be treated to something out of the ordinary was when a frosted cake was awarded for the cleanest dorm. The competition was held every Saturday and involved a lot of work. We dusted our rooms and washed our floors, waxing and polishing the tiles with a big machine that a student would sit on to increase the pressure and ensure the shiniest floor. We so wanted to win that cake.

I wasn't the only one who had positive experiences in Churchill. Most of the Inuit students seemed to adapt quickly to the new system and to excel in the new world order. The Churchill gang, for the most part, seem to have fared well. Many became high-profile leaders, such as Jose Kusugak, Eric Tagoona, and Mark R. Gordon (although we lost Mark far too young). My brother Charlie became a supervisor at the Churchill Vocational Centre, and later in his political career he was appointed by the retiring Pierre Elliott Trudeau as a Liberal senator, remaining one of the longest-standing senators in Canada today. Others like William Tagoona, Alec Gordon, and Rassi Nashalik went on to become well-known CBC journalists and broadcasters. Nancy Tupiq, my fellow dancer for the Harpoons, has been the assistant to the government of Nunavut's official clerk since Nunavut was created. My cousin Annie Watt was the director general of Nunavik's Kativik School Board for seventeen years.

Many others became leaders in their communities, holding positions of high rank within their organizations. Eva Aariak was the premier of Nunavut for a term. Michael Kusugak is a well-known author of children's books. Nancy Annirniliak is head of Parks Canada for Nunavut. John Amagoalik is known as the father of Nunavut because of his past political work on the creation of what is now the territory of Nunavut. Piita Irniq was the second commissioner for the government of Nunavut. The list goes on.

In hindsight, I suspect that what made the move from our traditional way of life on the ice and snow to this new schooling system work was the disciplined lifestyle. It resonated with us on several levels. We came from a highly disciplined culture, where we had to have self-control and respect for one another, as well as for nature, in order to survive. Our hunting way of life had passed down to us character-building skills that allowed us to adapt quickly to changes in our circumstances and our environment. To understand how we felt in Churchill, one also needs to grasp the importance of the feeling of safety we had. I have talked about how my childhood was colored by a sense of security because of the strength our families and our culture offered us. Churchill not only provided us with a familiar—although externally enforced—discipline but also gave most of us a sense of comfort from being all together in one place. I say "most of us" because I want to respect the fact that not everyone may have felt this way.

Yet although we may have felt relatively secure, many of us simply weren't as safe as we would have been in our own homes. The sad truth of that became clear during my second year in Churchill. My dormitory was away for the weekend at Camp Nanook—a small camp outside Churchill where students were taken to spend time in nature. While my group was gone, three teenaged boys from the hostel noticed polar bear tracks outside the Knight Hall gym. Having spent their childhoods hunting with their families, they were excited about the idea of tracking the bear. They set off, following the huge paw prints. The tracks disappeared around some large rocks a short distance from the gym. Two of the boys climbed up on the rocks to look for the bear. Paul Meeko stayed on the ground, following the prints around the corner. There he came face to face with the polar bear. The bear attacked. It would take a fatal bullet from an RCMP rifle to force the bear to drop Paul. He was raced to the hospital, but his injuries were so severe that he couldn't be saved. Before the arrival of the RCMP, all the students who were in the gym and many others had raced outside when they heard the commotion and became witnesses to the horrific scene. Everyone was traumatized and overwhelmed with grief. One of our own had been killed, by a polar bear at that. In our dorm rooms, far from home, we listened to Paul's girlfriend wailing with grief throughout the night.

There is no denying that accidents could happen in our home communities, but looking back at this particular tragedy, I'm struck by the fact that we Inuit children had been removed from the people who would have taught us life-saving skills about our Arctic wildlife. If Paul and his friends had been allowed to continue learning hunting skills from their elders, they would have known how to track a bear safely, and they would have learned to follow a polar bear only when necessary and only while armed. Instead, at Churchill, the boys were being taught welding, carpentry, and auto mechanics.

This was, of course, a time when many First Nations students attending residential schools were endangered by those who ran the institutions, and when a culture of silence surrounded this violation of children. Many of us from Churchill have talked about the fact that most of us never witnessed or experienced abuse—certainly nothing like what those attending mission-run residential schools at the time went through.

The history of residential schools in Canada spans nearly a century, with the last school closing as recently as 1996. During this time, the federal government, in an attempt to aggressively assimilate Aboriginal children, oversaw the schools, many of which were run by Christian churches. Approximately 150,000 children in all were taken from their families to be "re-educated" in English or French and Christianity. Resistance was rewarded with punishment, and many students experienced physical, emotional, and sexual abuse. Today, people are still trying to heal from these horrific experiences.

What's more, families were torn apart and unable to pass on tradition and culture when their children were abruptly removed from their care and from their communities. This has resulted in generations of trauma suffered by Aboriginal families across Canada. Prime Minister Harper offered an official apology for the residential-school system in 2008, a year after financial compensation was announced for survivors of the schools. The Truth and Reconciliation Commission (TRC) was established as part of a settlement when residential-school survivors and the Assembly of First Nations took the federal government to court in a class-action suit. The TRC worked to document what took place in residential schools and to provide a safe space in which survivors and their families could share their experiences through the medium of their choice.

I've lived most of my adult life coming to terms with my own intergenerational legacies, including being uprooted as a young child and losing my language and my connection to my culture, to my country food, and to my family, but even this doesn't compare to the horror that Aboriginal children went through in the mission-run schools during that time. So while most of us at the Churchill Vocational Centre were not aware of any abuses against students, recently, as a result of the work of the TRC, we have discovered that some former Churchill students have come to speak out about their traumas. There can be no doubt that when children are in vulnerable situations, such as the one we were in, without the supervision of parents and in the hands of strangers, there will be times when some of those students are not entirely safe.

Although it might not have been the majority who were abused at Churchill, unlike at the mission-run residential schools, even one abused child is one too many.

In all, I spent three years in Churchill. After students had attended the school for several years, the government came to interview them to decide who would then move on to Winnipeg or Ottawa to finish up their training at Algonquin College, if they were in vocational stream, or at high school, if they were in the academic stream. The official interviewed us individually and asked questions about what we wanted to do after high school and whether we wished to move to Ottawa or Winnipeg. I remember telling him I wanted to go into medicine and that I preferred Ottawa since my older sister, Bridget, was going to be there. We all understood that our stay at the hostel wouldn't normally exceed three or four years, so this was an expected move. I went home for the summer knowing that both my sister and I would be moving to Ottawa in the fall. My sister would attend Algonquin College, while I would start high school, along with my cousin Annie, my uncle Johnny's daughter. Lizzie had opted to stay in Churchill for a fourth year.

LIKE ANY TEENAGERS, we were excited to be going to a new city, and since I had already lived in the South, the buildings and landscape were not entirely foreign to me. What was different was actually living in a city and taking buses with Annie to visit my sister, Bridget, and my cousin Willie, Annie's older brother. Traveling on our own like this took some adjustment. We also were all placed with families whom we had never met. Annie and I were billeted at a home close to Gloucester High School. My sister was placed with a family nearer to Algonquin College.

Living with families I didn't know proved at times to be difficult, especially because I ended up with four different families in the three short years I spent in Ottawa. And while my sister, who had always been a quiet rock for me, was also in the city, it wasn't the same as living under one roof, as we had in Churchill. Although communicating with my mother back home became easier while living in Ottawa, she was still far away. Other than one visit while we were in Churchill, we didn't see her during those long ten months away from home.

The other challenge I faced in Ottawa was that the high school curriculum was proving to be difficult. I was weak in mathematics and physics, as well as French, which I had never taken before. These subjects started to lower my grade point average and, as a result, my confidence. My aim growing up, shaped by my mother's career and the Rosses' work, had been to become a nurse. At a certain point this goal evolved, and I decided I wanted to become Nunavik's first medical doctor. But if I couldn't excel in the sciences, this goal was at risk.

My love of sports waned, as well. I played basketball the first year I was in Ottawa, but I found it difficult to get to practices and games because I had no one to drive me or pick me up. I didn't continue after the first year. I had also developed the nasty habit of smoking cigarettes.

Unlike in Churchill, I got to know few people in the Ottawa high school. One *qallunaaq* friend, Kim Medveduke, introduced me to some school dances, but they weren't the same as our good old Churchill dances, where we all knew one another and had a sense of community. In Ottawa, most of my social gatherings were with fellow Inuit students scattered across the city. We would hang out at Sparks Street Mall, go to movies and rock concerts, or attend events organized by the counseling offices of the Vocational Training Department, headed by Ralph Ritcey, who was still in charge of Inuit students away from home. I yearned for the comfort of my fellow Inuit, who wouldn't comment on my "white" appearance, or ask where my parents or siblings were.

Ottawa wasn't a bad experience, per se, and life there certainly opened up my horizons in many ways. The first family we lived with were Mennonites, Jake and Ruth Enns, and both had had northern experience, having lived in Iqaluit, then called Frobisher Bay. We went to church with them and for the first time in our lives were part of a church service in which people dressed in everyday clothes and played the guitar while singing hymns. It was a striking contrast to our traditional Anglican church, with its robed ministers; long, dry services; and formal behavior. We thought the Mennonite services were quite interesting—and cool—and we would talk about them to our fellow Inuit students in Ottawa.

The Mennonite family were also pacifists, and with the Vietnam War raging, they took in a couple of American draft dodgers. These men were both attending university, and I believe one of them was a medical student. The Americans were kind to us, and since one of them had a car, he would even drive us places when we needed rides to visit friends or go to events. In getting to know them, we learned how difficult it was for those young students to be away from home, in a situation even more trying than the one we were in—

hiding from their own government so they wouldn't have to go to war and face the possibility of being killed. I remember thinking Canada must be a "good guy" to be taking them in. I've often wondered over the years what happened to those young men after the war. Years down the road, when President Carter pardoned the draft dodgers, they came immediately to mind and heart.

While my cousin Annie and I lived together the first year, we were separated halfway through our second year in Ottawa. We were moved into two different households because the first family's situation had changed, and they could no longer host two Inuit children. This went on year after year, and the inconsistency and lack of a familial safety net wore on me. One family I was placed with had obviously taken me in because the single mom was trying to make ends meet, and the government would pay her to have me in her house. During the week, she would turn off the hot water, and I wouldn't be able to shower or do my laundry until the weekend came around. She had given me her daughter's room, and her daughter stayed in her room. The mother took over the living room. I had only my room, the bathroom, and kitchen to occupy, tiptoeing between the three anytime I was in her house. It wasn't a joyful situation for me. My last year of high school in Ottawa proved to be happier, as I moved in with a family that had had a connection to my immediate family in Fort Chimo. During the previous year, I had become good friends with Ann Schulz and Patricia Ireland, who lived just down the street. We would walk together across the field to get to Hillcrest High School, and I got to know them well. I visited Ann's place often. One day I learned that her father had been to Fort Chimo many years before for some sort of technical work. He started to show me some pictures he'd taken while working in my hometown.

You can imagine how surprised I was when he showed me a photo of my grandmother, mother, and uncle Johnny. He was just as surprised. We were all thrilled by this connection, and the following year, when Ann's brother Derek went off on a year's trek through Europe, the Schulzes asked if I would like to live with them. That year proved to be a memorable time of friendships and caring from Ann's mother, who was kind and sweet to me.

It also allowed me to share with the rest of Canada one of our most powerful experiences as a nation. Like so many other Canadians, I clearly remember watching the famous 1972 Canada–Russia hockey game. Sitting next door at the Irelands' with Patricia's family, in front of the living room TV, I cheered along with everyone else when Paul Henderson scored Canada's winning goal.

While moving in with my friend's family made my final year in Ottawa more positive, the inconsistencies over the years in my home life meant that my academic hardships were compounded. I continued to have difficulty with chemistry, math, and physics. I knew these were important prerequisites for entrance into medicine. My hopes were further tarnished when one of the government counselors said that my marks weren't high enough to be accepted into medical school. He said that I was aiming too high and suggested I change my plans and become a nurse's assistant instead. After visiting him in his office downtown, my sixteen- or seventeen-year-old self stepped into the elevator and disparagingly thought, "If I'm not going to become a doctor, I just won't become anything." I didn't know how to grapple with this loss of direction.

I felt crushed by the idea that I wasn't going to reach my goal of becoming a doctor, the means through which I had thought I would best be able to serve my community. I wanted to follow in my mother's footsteps, after seeing her great service to our people. I had modeled my goals after the example she set, and they had been dashed. In hindsight, I have often wondered why tutoring wasn't offered to help me raise the marks in my weaker subjects. Instead, it seemed easier for the counselors to offer, with their discouraging remarks, a less ambitious route for me.

My life, goals, and family situation were changing after three years in Ottawa. My sister, Bridget, had gone to Greenland as an exchange student. She had fallen in love with a Greenlander and was now expecting a baby. Knowing how my mother had raised us, so clearly not wanting her daughters to repeat the legacy of fatherlessness for her grandchildren, I knew it would be difficult for her to handle this news. My mother was also suffering from a heart condition, and since my goal was no longer being fulfilled in Ottawa, I felt I needed to return to my family and my culture, to support my mother and sister and recalibrate myself. In the end, the father, Tikili, did immigrate to Canada, and he and Bridget married and had three children together.

After spending eight years in southern Canada, raised by strangers; going to school far away from home; being disconnected from family, culture, traditions, and community; and losing so much of those formative developmental years, at eighteen, I decided it was time to go home. I left Ottawa and traveled back to Kuujjuaq.

My return to the North was the right move for me, and yet I wasn't entirely prepared for the changes that would greet me upon my return. The safe, supportive, strong community I remembered so well from my early childhood was being battered and bruised. And a way of life that I remembered fondly was slipping away. But perhaps I shouldn't have been so surprised. I had already been given glimpses of the changed Arctic during my short visits home. The most dramatic had occurred during my only winter return in my time down south.

My first year in Ottawa, I had asked my federal counselor, "If my mother pays half of my ticket, can I go home for Christmas?" He said yes, and my mother agreed to share the cost of a flight to Kuujjuaq. I was thrilled to return to an Arctic winter, to the ice and snow of my childhood. When I got home, however, the changes were stark.

The first afternoon after my return, I left the house to walk over to the HBC store. As I walked along the snow-packed road, I was surprised by a burst of sound, a roar so loud it might have come from a jet plane taking off right beside me. I jumped and turned to see a snowmobile blazing past me. It was then I realized that I hadn't seen any dog teams. As my Christmas visit unfolded, I would learn that my brother Elijah's dogs were gone, as were those of my uncle Johnny. I was so preoccupied with avoiding the new noisy machines (even asking my mother to walk with me to events at night so I wouldn't be run over by them), as well as with my teenage priorities, that I failed to really question what had happened to the dogs. In fact, it wasn't until I was working as an adult at Makivik Corporation that the story started to unfold. So horrific was this story, and the wounds caused by it so deep, that no one spoke about it for years. But as I would discover, it was just one of many tragedies to befall my community.

3

A Return Home

I RETURNED TO KUUJJUAQ at eighteen. My mother was now semi-retired, the nursing station all but shut down after the new Quebec health system took over. My sister, who was expecting her first-born, was living with us, although soon she would move next door when Tikili arrived from Greenland and they married. Elijah and his family also lived next door, and Charlie and Ida were just around the corner. But despite being surrounded by family, I felt an absence in my mother's house. My many summers at home without my grandmother had not gotten me used to the hole that her death created.

A short time after returning, I started to work as an interpreter at the nursing station clinic across from my mother's house. This was a bit ironic as I had lost so much of my Inuktitut while living in the South, a fact that I was very aware of. Over the years, I had been flooded with feelings of inadequacy when I struggled to express myself in my mother tongue. But I still forced myself to speak, even when I made mistakes, as most of the older generation didn't speak any English. (To this day, most Nunavik communities speak more Inuktitut than English. In fact, the language is vibrant in the area, with the young people, as well as the elders, being proficient in it.) For Lizzie and me, to lose Inuktitut was to lose the ability to communicate easily with the majority of our family and community. Lizzie was especially affected by the loss of Inuktitut because her parents spoke no English at all. Lizzie's older siblings had to interpret for her when she wanted to communicate with her mother or father.

When Lizzie and I made mistakes or had difficulty explaining ourselves, some people laughed. I think this affected Lizzie more than it did me—she didn't speak Inuktitut publicly for at least fifteen years (although she spoke it with her children when she became a mother). But many people in the community also helped us when we were at a loss for words or failed to express ourselves clearly. And I continued to try to speak the language I had grown up with.

In fact, my first summer job, the year I turned sixteen, was as an interpreter at the small hospital across the street from my mother's house. It was one of the recently built, modern two-story buildings and housed a clinic, a laboratory, and an X-ray room. On the second floor was also a small hospital ward and the Direction générale du Nouveau-Québec (DGNQ), the Quebec provincial offices. There was also a cafeteria where DGNQ employees would have their meals, and where the clinic staff could take coffee breaks.

I started in the hygiene department as a summer student working with an older French-Canadian nurse named Evelyn Bossé. She taught me how to give vaccinations, help her with her examinations of expectant mothers, and deal with ailments such as impetigo and scabies, both highly infectious skin conditions that a lack of hygiene is one of the causes. We were still living in small homes without running water in those days, and keeping clean was a real challenge for some larger families. I remember having to bathe many members of certain families at the clinic, so they wouldn't spread these conditions to other family members or the community. And, of course, I also interpreted for the nurse and the patients.

When I finally came home to stay at eighteen, I started working at the same hospital building but moved to another clinic on the same floor as the hygiene clinic. In a way it was a promotion for me—now I was working with the doctors. At first the job proved to be challenging, as I had to search for so many words in Inuktitut. The understanding of the patients and others who knew me was helpful, however, and I did manage to become the communicator between the doctor and patient. Relearning Inuktitut in this way started to give me back my sense of Inuit identity. There's nothing worse than no longer being able to communicate effectively in your mother tongue with your own family and community members.

I felt elated by my growing comfort with Inuktitut and loved this work that connected me with the entire community. And I think the people I served also felt comforted having me there, just as they had when my mother was their main interpreter many years before. My favorite times were when the elders would come in. They were grateful and loving to me, perhaps especially so because they had all been close to my grandmother. Their kindness filled that void my grandmother had left behind, and their gentle energy was what I had so missed and longed for since my grandmother's passing.

In many ways, I was able to act as a pseudo–doctor's assistant. Until the mid-1970s, the number of employees at the clinic was limited. That meant that a great deal of flexibility was allowed in each position. As an interpreter, I was trained on the X-ray machine and occasionally traveled with the doctors on their visits to other Nunavik communities. I learned a lot about medicine and even helped to deliver eight babies.

Along with treating patients from Kuujjuaq, the staff at the clinic would see patients from other communities who had flown in to our regional hospital and clinic if the nursing stations in their own regions couldn't treat or attend to them. Women, in particular, would come to Kuujjuaq to deliver their babies. The staff at the clinic were respectful of the wishes of the mother, particularly about whom she wanted in the room and whom she wished to cut the umbilical cord. For this reason, sometimes I'd be asked to attend a birth if the mother was a friend or related to me in some way. At other times, I happened to be working with a particular doctor who was going to do the delivery that day. With the mother's permission, I would also be part of this miraculous experience.

In our culture, the person who cuts a baby's umbilical cord plays a special role in the life of that baby. That person is forever known as the *sanajik* of the baby when it's a girl, or the *arnaqutik* when it's a boy. The baby then becomes the *arnaliaq* of the *sanajik* or the *angusiaq* of the *arnaqutik*. An Inuk girl will give the first thing that she has crafted to her *sanajik*. An Inuk boy gives his first hunt to his *arnaqutik*. This is considered a rite of passage for the young man, and rituals and cultural significance are associated with the delivery of this first hunt.

My own grandmother had several *angusiaq*s and *arnaliaq*s. The *angusiaq*s would bring her their first hunt, and a remarkable ritual would follow. As the young man watched, my grandmother, his *arnaqutik*, would break into ritualistic movements, imitating the animal by rolling on the floor and making all sorts of strange movements and noises. She would also pretend to bite the hands of the hunter as a way to acknowledge his ability and the power of his hands and to encourage him to be a great hunter. My grandmother was usually a calm and collected Inuk woman, so standing on the side and watching her become this different dynamic personality made us, her grandchildren, somewhat shy and a little embarrassed. Yet we understood that this ritual was a necessary part of our hunting culture and tradition.

The rituals with which a hunt or first craft were received were designed to recognize the accomplishments of each *angusiaq* and *arnaliaq*, as well as to express gratitude for the harvest or creation that brought food or something useful to her. At the end of the ritual, the young men and women emerged with higher levels of confidence, knowing that their hunt or their work had been affirmed, validated, and valued. They never forgot this moment.

During her time working in the nursing station, my mother came to have many *arnaliaqs* and *angusiaqs* herself, and in fact she once delivered a baby on her own in Kuujjuarapik, where she was stationed for a time after the federal nursing station closed down in Kuujjuaq. So I certainly grew up appreciating the meaning and importance of this tradition that helped to build confidence in young people.

We have lost many of these practices, but I have tried to keep up with the lives of my own *angusiaqs* and *arnaliaqs*. I have witnessed the journey of these young people as they have grown to become great hunters and providers for their families and community. Some have grown up to become accomplished entrepreneurs and political figures in the community, and they are a source of pride for me, their *arnaqutik* and *sanajik*. Many of the men have brought me their first hunt, and while I didn't perform all the movements and chanting that my grandmother did, I expressed a sense of pride in the hunter and in keeping this tradition alive. And I felt great to be honored by these young men and women. These are powerful memories for me.

I truly treasured being allowed to participate in the eight births I attended. I also valued the ties with the extended Nunavik community and with the doctors who served it. In particular, Dr. Normand Tremblay, who worked in the Nunavik health sector for many years, strongly influenced my love of the work. He involved me as much as he could in many of his medical interventions. He also had a great sense of humor, which helped us to get through some challenging times. I recall, as well, flying into several communities in Nunavik with Dr. Bruno Dumas. Dr. Dumas appreciated my interest in medicine and would involve me when he could in tending to patients. I also worked closely with a third doctor, Dr. Laurent Letourneau. Laurent was a highly strung man who, interestingly enough, seemed to rely on my calm demeanour to get through many of his busy days. He could get overwhelmed easily or overreact to certain situations. He found having to write up patient files at the end of the day trying, and so I'd stand beside him, reminding him what the patient came for and what medication he'd prescribed and calm him down as he finished his paperwork. All these doctors were amazing teachers and became my close friends (Laurent would deliver my first child) and had an enormous influence on me during my early adulthood.

I LOVED MY WORK at the clinic. I felt useful, and it was the next best thing to being a doctor. It also proved to be meaningful for me, as I both witnessed and was a part of the rapid changes that were affecting my community and the North.

While coming home after eight years in the South was a

relief, it was also something of a revelation. I noticed that the community seemed different somehow. At first, the changes were difficult to define. But my work in the clinic gave me a first-hand view of some of the problems that were emerging.

Although the Kuujjuaq hospital and clinic were small, they were always busy, with health issues escalating in our communities over the years. The clinic was seeing a lot of respiratory illnesses in small babies and children. This was no doubt due to the conditions in the tiny frame and matchbox houses that had been provided to so many families in Kuujjuaq. Constructed to retain heat in the cold, with poor ventilation, these homes often harbored mold and bacteria. And as so many people smoked in those days, they were often filled with smoke.

The crowding and lack of running water in these homes also led to the rapid spread of viral and bacterial infections. Gastrointestinal ailments, skin infections, chicken pox, some strains of measles, and hepatitis hit our communities hard in those years. A number of epidemics spread through our community while I was working at the clinic. But most troubling of all was the number of patients we were seeing with severe physical traumas.

As I have said, my overwhelming sense during my first ten years in Old and New Fort Chimo was of safety and security. But that's not to say that the community was without problems. As in any community, there were a few people whom we children felt unsafe to be around. But during my summer visits, I had noticed with each passing year that more and more episodes of erratic behavior, public drunkenness, and brawling seemed to occur. When I started working full time at the clinic, I saw clear evidence that alcohol and drug addictions were on the rise.

I was alarmed to see that some people were coming to the clinic and the hospital with injuries resulting from drunk driving. Even more disturbing was the number of victims of violence we saw, victims of fights and abuse that were clearly the result of intoxication. At the end of the week, supply planes would land with the beer and liquor everyone had ordered. Then the weekend binge drinking would start, and so would the incidents of public drunkenness and abuse. Most violence happened within the family.

People whom I had known all my life were coming into the clinic with horrific injuries, like broken bones and gashes requiring stitches. Sometimes the injuries were so severe, the patients needed hospitalization. Some were beyond our hospital's ability to treat and would require transport to Montreal. One man had to be medevaced south after his brother broke his skull with a sink he'd ripped off the wall. These were men I had grown up with, but their behaviors now put them in stark opposition to the decent and compassionate men we knew them to be.

It became common to see many in the community with black eyes and bruises after a weekend of drinking. Parties replaced weekend hunting and fishing trips for some people. (However, the camping and hunting weekends would come back in the next couple of decades as most members of the community who'd taken up partying went back to the cultural activities as they aged.)

As a young woman, I was witnessing the painful, destructive connection between addiction and alcohol abuse and violence. And I was seeing that it was women who were bearing the brunt of this terrible social unraveling.

Since I was the X-ray technician, as well as the interpreter, I was often called in when an injured woman showed up at the clinic. One day I arrived to take an X-ray and found an older woman whom I knew well lying on a stretcher. She was in excruciating pain, with many injuries and a mass of bruises. She had, apparently, been out camping on the weekend with her husband. He had brought alcohol along. At some point, in a drunken rage, he had beaten her with a tree. Eventually word got back to Kuujjuaq, and a helicopter was sent to retrieve this remarkably gentle woman. By this time, I had seen a number of severely injured people, so I was not as shaken by her wounds as I might have been. I tried to be gentle and didn't say much, hoping to avoid making her feel embarrassed. But pain and shame were in her eyes.

During my four years at the clinic, I saw many women of my community, gentle women of integrity and dignity, in a similar state. I could sense that they were embarrassed to have me at such close range to their trauma. They acted almost apologetic that I had to bear witness to their suffering. I tried always to be tactful and kind, and to remember that the husbands who had inflicted such pain on the ones they loved were, when sober, often great providers, fathers, and community members. In fact, some of those who were most violent when drunk were extremely caring and accomplished men. Without alcohol, very few, if any, of these men were violent to their family members, or to anyone else for that matter.

But it was hard to reconcile the men we knew, the neighbors, fathers, uncles, brothers, with the things they were doing under the influence of alcohol. For the victims, it was no doubt an especially confusing and paralyzing situation.

I share these stories not to be sensationalistic or to judge those involved, but rather to show how quickly in my own life, in less than twenty years, the tumultuous changes from the outside were affecting the very core and soul of the grounded, reflective, caring hunter spirit of our men.

In the past, we had seen our men working on their sleds, carefully, meticulously preparing the runners; tightening the ropes; focusing on the tiniest details. And we had seen them heading out onto the land. We all knew they would be spending hours, days in silence and stillness, disappearing into the landscape so that caribou or the seals might appear. Their masterful control and focus were more important than anything. But later, as the years passed, we started to see men traveling out onto the land with booze packed alongside their food and supplies in the *qamutiik*. How could they hunt when they were drinking? How could they afford to engage in reckless habits in a remote and often unforgiving terrain, where clarity and focus are a must and where injury might quickly lead to death? How could they accept these bursts of violence when we as Inuit have traditionally looked at anger and loss of control as the most childish of behaviors? And why would they sacrifice the meditative, clarifying, healing experience of being on the land in exchange for mind-fogging booze?

We didn't understand at the time that this contradiction in terms pointed to something other than individual weakness. It was evidence of a breakdown of Inuit society. It would be years, perhaps decades, before we began to talk about something that I call "the wounded hunter spirit." Years of pent-up anger and frustration caused by the tumultuous changes our people had experienced were finding an outlet in alcohol abuse, addiction, and violence. Decades later, I began to appreciate the historical traumas that had led to this damaged state.

Perhaps the first of these great upheavals was the switch from our hunting-and-fishing culture to one of trapping and trading. When the Hudson's Bay Company arrived in the Arctic, it brought with it the industry of fur trading and trapping. Inuit went from being traditional hunters to trappers in one generation. Meeting market demands for furs brought us a new way of life. Since the focus of hunting was now divided between food and furs, our diet became more dependent on southern food, which was both limited and difficult to attain except through barter, the system established by the HBC. Our hunters also switched from traditional hunting weapons to rifles and ammunition, which they acquired through trade with the Hudson's Bay Company. We had no choice but to become more, in the absence of a better word, southernized.

The integration of Inuit into the trapping economy changed us from self-sufficient hunters for food to trappers for fur. Trying to do both wasn't practical, especially when hunters needed to use what animals they did harvest from the land to feed their dog teams. The less we hunted for our own tables to serve the larger global market for fur, the less nutrition we were getting from our country food. This was the beginning of a change in our diet that would eventually lead to poorer health. More harmful still, however, was the collapse of fox fur prices in the late 1940s. When the prices fell, trapping became unsustainable for many Inuit families. Government assistance programs, primarily welfare and family allowances, started to become primary sources of income. But the people living in outpost camps or those who had no fixed address couldn't collect these payments. So now our once semi-nomadic people had to move into permanent settlements, often far from their traditional hunting grounds.

Our populations became increasingly dependent on the government and southern institutions for survival. As Frank Tester of the University of British Columbia and Peter Kulchyski of Trent University say in *Tammarniit (Mistakes)*, their penetrating book on the effects of government policy in the eastern Arctic communities, "The role of the churches, the Hudson's Bay Company and other traders in creating conditions of dependency was increasingly the subject of controversy and debate within the Arctic administration as the crisis intensified following the Second World War." The government-initiated family allowances also played a large part in the moving of Inuit into settlement living. The Family Allowances Act contained a provision that required mandatory school attendance for children. The term "schooling," in relation to First Nations and Inuit children, however, could include "training in traditional hunting, trapping, sewing, or other pursuits where no possibility existed of attending school." Yet this provision was generally ignored. Schools had been built in order to educate Inuit children; it was now government policy to ensure that those schools were filled with students. To do so, the government withheld family allowances unless families moved off the land, into the communities, and sent their children to school.

While undoubtedly it was easier for the government to promote assimilation and to provide services, including schooling, to a population living in settlements, there was another likely reason that some Inuit were coerced into moving off the land. In the 1950s a number of Inuit families were forcibly moved from Arctic Quebec to the High Arctic to form settlements in such places as Resolute Bay and Grise Fiord. The families weren't used to the much harsher climate of the Far North, and survival was a struggle. In fact, wildlife and other food sources were so scarce, some families had to scavenge from the RCMP garbage for sustenance, according to John Amagoalik, whose family was part of the government move. Many were deeply traumatized by being torn from their familiar homeland and relocated to this extremely isolated, difficult environment, and this pain continues to be felt by the now-adult children who experienced these moves—and by subsequent generations. (In her sobering 2009 National Film Board documentary, *Martha of the North*, Martha Flaherty captures the horror and sorrow of the forced relocation.) While the government claimed to be attempting to improve the living conditions of the transplanted Inuit, many have asserted that Ottawa created the settlements so that it could stake a territorial right to the High Arctic. Whatever the government's motives, its refiguring of the Inuit communities was effective. By 1965 most Inuit were living in settlements.

But settlements wouldn't necessarily provide a better life or more security for our communities. This was clear as early as the 1940s, when the fur trade disappeared. By the time the Second World War was in full swing, many in the Arctic, including the families around Fort Chimo, were facing starvation. Without a trapping income, families simply couldn't come up with the money to buy food and supplies—or to buy ammunition that would allow them to hunt. The situation was desperate, and yet the Canadian government, preoccupied by the war effort, seemed to have utterly forgotten about us.

Then the Americans arrived. They had come to build the airstrips that would be their stopover points on the way to Europe. But they also brought with them food, supplies, and jobs for our men and women. They literally saved many in and around our little community of Fort Chimo, as well as a number of other northern areas.

Their presence was still felt when I was child, and it went beyond the physical tracks and barrels they left behind. My mother and grandmother talked about the arrival of the

Americans when food was scarce and their world was rife with hunger. My mother always said that they wouldn't have made it without the Americans. My uncle Johnny, who is now eighty-six, also talks about the starvation and disease they endured in the forties, noting that Inuit of my region were more dependent on trapping than any others in the eastern Arctic, and so were more strongly affected by the disappearance of the trade. Without the family allowances that had recently been introduced, the deprivation and starvation would have been much worse. But it was the arrival of the Americans, with their food and their jobs, that saved so many in Fort Chimo, and in other parts of Nunavik and Nunavut.

Because of that early relationship, we Inuit have remained attuned to what happens in the United States. I remember when John F. Kennedy was assassinated. The elders and adults in the community talked about the tragedy for days. And on September 11, 2001, it was clear that this connection was still felt in the Arctic. Our airwaves were brimming with memories and stories of how the Americans had arrived in the 1940s. We held America in our hearts and our minds, and felt great sorrow for their loss.

I was reminded again of the important place America has in the hearts of many Inuit in 2009, when I was about to head to Bowdoin College in Brunswick, Maine, as a visiting scholar. Before I left Kuujjuaq, where I had been visiting, my uncle Johnny said, "Thank the Americans for us. Thank them for what they did for us during the forties when we were going through hunger." I was touched by his sentiment and his sincerity, and wondered how long he had been waiting for the opportunity to express his gratitude. With deep compassion and respect for my uncle, I relayed his message to my class at Bowdoin. This thank you, conveyed to a small class of American students from an elder Inuk man, was meaningful, symbolic, and potent for all of us.

While forced resettlement in the forties, fifties, and sixties was a blow to our traditional way of life, our hunters would be dealt another strike in the sixties, when Brigitte Bardot came to North America to campaign against commercial harp seal hunting in Newfoundland. She and other animal rights activists were quite successful in their fight against the hunt, and the market for sealskin products collapsed almost overnight. Even though these activists didn't target the Inuit traditional hunt of seals, their actions destroyed our ability to sell the by-product of our subsistence hunt. Inuit hunters, who could make as much as eighteen thousand dollars a year selling surplus pelts, saw their income dwindle to nothing overnight. Our families became increasingly dependent on government support. And our hunters' pride and self-worth were damaged anew.

There was also, of course, the trauma parents and children suffered when the children were sent away to school, as I was, and when family members were sent south for medical reasons. After the Second World War, a tuberculosis epidemic in the Arctic sent many Inuit to southern sanatoriums. Other Inuit were sent south suffering from flu, polio, or typhus. Many of these people were never heard from or seen again, presumably having died in unfamiliar settings in southern hospitals. Not knowing the fate of their loved ones, many today still search for their family members' burial places. And then there was the loss of the sled dogs.

It was many years after I had come home, at fifteen, for Christmas and noticed all the noisy snowmobiles that I finally began to learn about the appearance of the machines—and the disappearance of the dogs that preceded them.

It took me decades to discover this truth about our dogs.

Silence descended upon our hunters and the community. Few people were willing to discuss it—not even my own brothers and uncle. The wounds were so deep, the events so shocking. But there was also, I suspect, a sense of shame about what happened and about our community's failure to stop the systematic slaughter of their dogs.

Two films have documented this tragic history: *Qimmit: A Clash of Two Truths*, produced by the National Film Board and Piksuk Media in Nunavut, and *Echo of the Last Howl*, produced by Taqramiut Nipingat Inc. and the Makivik Corporation. According to these films and to Inuit testimony, RCMP officers and government officials traveled throughout the Baffin Region in the 1950s and 1960s searching out dog teams. Claiming that a number of dogs had been infected with canine distemper, and that some of the sick animals had attacked people, the officials took some dogs and sent them south for "health care." The dogs never returned. More often, however, they instructed hunters to bring their dogs to a designated spot. The animals were not inspected for illness, no questions were posed about their health or behavior. Certainly no permission was asked of the owners. The dogs were simply shot. In some instances, the carcasses were thrown in piles and burned. All this happened in view of their shocked owners. For the Nunavut perspectives of the dog slaughters, see the Qikiqtani Truth Commission Reports, which were commissioned by the Qikiqtani Inuit Association.

The testimony of Inuit who watched the slaughter unfold is harrowing. Some men had come in from outpost camps and watched as their only means of transport, their only way to get back to their families, was destroyed before their eyes. Others said that they were preparing to go hunting, and their dogs were shot and killed as they stood harnessed to the sleds. Still others testified that the RCMP chased and shot loose dogs, even firing at those that had taken refuge under family homes. Some dogs were wounded and not killed, and their owners would beg the officials to track the animals down to put them out of their suffering. My own uncle Johnny eventually told me that he received a knock on his door, only to have someone of authority throw his new harnesses in his face and tell him, without remorse or apology, that he had just shot his dogs. And children, many of whom had been assigned a particular dog to tend from their family's team, watched helplessly as the animals in their care were shot. In all, over twelve hundred dogs were destroyed. And while the official explanation given at the time was that they were culled to prevent the spread of distemper and attacks by sick dogs, many now suspect that the destruction of the dog teams was another way to force Inuit families to move from outpost camps into settlements by removing their only mode of transportation.

Anyone who has ever cared for an animal can understand the horror of this slaughter. While we are hunters, we kill animals only for food. Senseless killing is not accepted in our culture. What's more, the importance of the sled dogs to Inuit, in particular to our hunters, can't be overstated. Dogs were literally lifesavers for us. Not only did they lead the hunters out across the land, but the dogs' excellent sense of direction meant that they could find their way home in a snowstorm or blizzard. When hunters traveled by dog team, they rarely got lost. Dogs have been known to pull hunters who fell through the ice out of the freezing water; they have also been known to uncover buried snow houses after a blizzard, saving people from suffocation and starvation. And a member of a dog team might even serve as food in the rare cases where a storm or other calamity stranded a hunter.

In *Echo of the Last Howl*, Quitsaq Taqriasuk-Ivujivik describes the importance of dogs to an Inuk hunter's sense of self: "If a man owned dogs, he was a very self-sufficient provider because they made it possible for him to travel and hunt for his family. In other words, the dog teams made a man. Even if it seemed like a boy wouldn't become a real man, as soon as he started owning a dog team, he became a great hunter and a real man. He even ended up helping his neighbors by providing transportation and food."

Numerous families who lost their teams couldn't hunt in the winter for many seasons, creating great suffering and dependence on social assistance. While some were eventually able to replace their teams with snowmobiles, many never recovered financially or emotionally. A number of these men slipped into alcoholism.

It may be hard to understand why our Inuit men and women didn't protest the cold-hearted slaughter of the dogs that they not only depended on but also had a deep bond with. Just as one might wonder why parents allowed their children to be sent so far away for school, or why families would agree to move into settlements instead of living the way they had for generations. Did they really think that the *qallunaat* and the Southerners knew best? That all these things were the right thing to do? Of course not. To understand why our people followed directions that were clearly counter to their culture, their wisdom, and their own self-interest, one needs to understand what we Inuit call *ilira*.

In his book *The Other Side of Eden*, British anthropologist Hugh Brody translates this powerful Inuktitut word into English, describing it as "the mix of apprehension and fear that causes a suppression of opinion and voice." *Ilira*, he explains, is caused by "people or things that have power over you and can neither be controlled nor predicted. People or things that make you feel vulnerable, and to which you *are* vulnerable."

In those early days of interaction with *qallunaat* men— church missionaries, Hudson's Bay Company officials, and government representatives—our people often felt intimidated. The *qallunaat* had tools and technology that were foreign to us. And they clearly had power and a willingness to exercise it. What's more, their behavior was often at odds with our quiet, restrained way of approaching others. The *qallunaat's* overt displays of anger and frustration were unnerving, scary.

In *Echo of the Last Howl*, one hunter describes feelings of *ilira*, saying, "That morning we were told to take the dogs up to the bay. Compliant or submissive, worried, surprised, we obeyed, not daring to question." And Eli Qumaaluk puts it this way: "We had no choice but to relent. We would have felt ashamed to be arrested by the police for not complying."

Ilira may explain our acquiescence with the *qallunaat's* demands, but it doesn't capture the shame and pain caused by following directions that went counter to our culture. Poverty, dependency, coerced relocations, forced separation of families, the dog slaughter—the litany of historical traumas left many in our society angry and deeply troubled. The independence we had given up as a wise people of great ingenuity had been diminished by accepting too easily the new "bosses" who now had full control of our lives. These wounds hit our men harder than our women, but we all suffered as a result. The shame, the guilt, the loss of integrity and pride was turned inward and festered as anger and resentment. The calm, reflective, and wise hunter spirit was now being replaced by a wounded hunter who no longer felt at home in the changing world around him. While many Inuit women found work in the new world, transitioning from their role as caregiver to positions as interpreters or translators, teachers, or domestic helpers in the new institutions, many of our men struggled to find work. The hunter-provider was now being displaced by providers from the institutional world, like welfare or family allowances.

Of course, compounding this problem was the availability of alcohol. It's interesting to note that before our contact with Europeans, we Inuit had no mind-altering substances whatsoever. None that I'm aware of. But now a powerful one had taken hold of many of our people. Alcohol allowed these wounded hunters to release their feelings in a terribly destructive way.

It wasn't just the men or the older generation, however, who were showing signs of trauma during my young adulthood. During my first year back in Kuujjuaq, I heard for the first time of suicide. Although other suicides may have occurred in our community in the past, this was the first that most of us in Kuujjuaq knew of. The young woman who had taken her life was originally from Nunavut but had been living in Kuujjuaq with her boyfriend. She was a bit older than I was, and I had known her from our Churchill days. News of her death spread through the community at lightning speed. Within hours, everyone in Kuujjuaq knew of it.

The shock of this death was huge. It was beyond our comprehension. Suicide was a new animal, something we hadn't thought of before, hadn't considered as a possibility. It took a long time for us to accept that it had happened, and it left us with a feeling of vulnerability that we hadn't known before. Sadly, this tragedy would become all too familiar as the years passed.

AT ABOUT THIS TIME, Dr. Tremblay was involved in discussions with Laval University in Quebec City about starting up a new department within the university to serve the "North" (meaning northern Quebec). Dr. Yves Morin, who had been Dr. Tremblay's mentor during his medical training, and who was dean of the Faculty of Medicine, was pushing for this initiative. At the same time, the Centre d'études nordiques at Laval was also vying to be the department in charge of the North. They wanted to follow the same route that McGill University had taken with its teacher-training programs for Northwest Territories.

Dr. Tremblay was aware that the sweeping changes I was seeing in my community had made me especially passionate about my work at the clinic. He also knew that I had once hoped to be a doctor before my high school experience made this untenable. With this in mind, he proposed to Laval that I become one of the first candidates for a medical-training program created specifically for Inuit of Nunavik.

According to Dr. Tremblay, the university would have been open to this idea, providing I came back to work in Kuujjuaq. I must have shared the fact that I was "recovering" from eight years living away, and that being gone for another seven years to finish medical school seemed unrealistic for me, as it would have been for most Inuit in those days. Dr. Tremblay said the university would likely design something shorter, a more condensed course, for Inuit and for northern realities.

Since that time, other specially created programs for Inuit teachers and student counselors that allow participants to stay in Nunavik during their studies have been established by the Kativik School Board in partnership with McGill. But when I was eighteen or nineteen, those opportunities were few and far between. Laval's proposal was appealing. But before Dr. Tremblay and Laval University had gone any further in their discussions about this new department and the possibility of creating a program tailored for me and other Inuit candidates, I decided to take another direction.

During the early days of my return to Kuujjuaq, I had agreed to join my brother Charlie on a trip to Nain, Labrador, where, as president of the newly created Northern Quebec Inuit Association, Charlie was going to meet with the community and local leaders. He had invited me along as the cameraperson to film those meetings. At the Kuujjuaq airport, I met a French-Canadian dispatcher, Denis Cloutier, who was working for St. Felicien Air Services and living in Kuujjuaq. While our exchange was brief (essentially just me correcting him when he referred to Charlie as my husband), I continued to get to know him through our social circle. Eventually we started dating. We dated for a year or so, before marrying in the spring of 1974, when I was twenty. That put any plans to continue my education on the back burner.

I continued to work at the clinic, and Denis eventually got a job with Transport Canada as assistant to the airport manager in Kuujjuaq. As a young married couple, we were given housing with Transport Canada, and I moved out of my mother's home and into a house with running water for the first time in my hometown. My cousin Annie also married a French Canadian, Claude Grenier, later that summer in August. Both my cousin and I were married in the Anglican church, becoming part of a trend that saw *qallunaat* men actually staying and marrying Inuit women.

Denis and I stayed in Kuujjuaq, where I gave birth to my daughter, Sylvia, at the age of twenty-two in the very hospital in which I worked. When Sylvia was still a baby, the three of us moved to Mont-Joli, near Rimouski in Quebec. Denis wanted to become a crash-and-rescue firefighter for Transport Canada, and the on-the-job training for the position was in Mont-

Joli. Denis did well in the program and remained a firefighter for many years, eventually becoming the chief at the Dorval Airport in Montreal.

We didn't stay in Mont-Joli for long, as I found it difficult to adjust to the small, isolated French community. Although I had picked up a considerable amount of French working at the clinic with the French-speaking nurses and doctors, I didn't feel confident enough to speak it publicly. No doubt a lot of my discomfort in Mont-Joli harkened back to the first time I was sent away. The feelings of being far from home, in a strange community, struggling in a foreign language were too reminiscent of my time in Blanche. What's more, my young husband worked many night shifts, so I spent a lot of time by myself, my only company my baby girl and the mice that came to visit our little mobile home in the fall. The mice absolutely terrified me. To relieve my loneliness, Denis would drive us five or six long hours to Montreal to visit his parents on his monthly six days off.

But the trips to Montreal did nothing to alleviate my homesickness, especially as I was now expecting our second child. I was lucky to have a neighbor who spoke English, a rarity in the small town of Mont-Joli, and her short visits were most welcome.

In 1977, about a year after we moved to Mont-Joli, we relocated to Montreal after Denis got a position at the larger Dorval Airport. By this time our son, Eric, had been born. We lived in Montreal for a few years, and we bought our first home in Saint-Eustache, just outside of Montreal. At this point, we decided that we would rent out our home and move back to the North. Denis had applied to become fire chief at the airport in Kuujjuaq and gotten the position.

I was so excited to be going back home with my family.

Denis and I wanted our children to learn about both worlds, his and mine. And I was determined that, unlike in my childhood, the children would stay with their parents. I recognized that the trauma of living in the South for me was caused not so much by the South, in and of itself, but rather by the sudden severing from my mother, grandmother, and siblings. I knew that my children needed the safety and love of their parents to adapt and flourish while navigating different environments, cultures, languages, and food. To this day, my children are comfortable in both worlds. They eat and enjoy country food like their mother and appreciate southern cuisine just as well. (In fact, my daughter is a great fusion cook and creates some mighty fine dishes with caribou, ptarmigan, or fish.) It was important for them, as it was for me, to go back north and have my children experience my culture and my community. So that is what we did.

While my husband worked at Kuujjuaq's airport, I volunteered, along with my cousin Annie Watt, to get some much-needed services for our community. Annie and I had noticed many women in Kuujjuaq were unable to work because of the lack of child-care services. We set up a committee made up of several women from our community, including Soré Moller, a Greenlandic Inuk who was now a resident of Kuujjuaq and who had experience in setting up a daycare in Greenland, and Betsy Forrest, an Inuk teacher originally from Kuujjuarapik, another Nunavik community. Jean Guy Bousquet and Adele Yassa from the Kativik regional government's Department of Economic Development were also instrumental in helping start up this project. We then got down to the nitty-gritty of finding a suitable building (luckily, the Quebec government had vacated a number of buildings in the community), becoming incorporated, acquiring funds, and hiring staff. Louise Woodrow and Vickie Gordon became the center's two senior staff. We also had a public radio contest to choose the name for the daycare center. My aunt Penina won with Iqitauvik, which translates as "the place where one receives hugs." (This name continues to be used for one of the three daycares that now function in Kuujjuaq.)

Not long after this project, Annie and I worked on building three playgrounds for the children of fast-growing Kuujjuaq. We did radio shows to inform the community of the rationale behind building the playgrounds and to garner support in putting up the equipment. Denis was heavily involved in directing this work, as he had gained experience in building a playground for the children of Transport Canada's employees, close to where we were housed.

In addition to my volunteer work, I had begun working for our regional school authority, the Kativik School Board (KSB) in Kuujjuaq.

The Kativik School Board became, for me, a workplace and a source of training. My work with the board would deepen my understanding of the challenges that my community, and its newest generations, faced. It would also introduce me to the larger institutional and political arena in which I would spend the rest of my career. And ultimately, in my struggles with this organization, I would discover the importance of making my voice heard.

The KSB was created through the James Bay and Northern Quebec Agreement to serve the Inuit communities of Nunavik (then northern Quebec). Given that its first director general was from the South, and that the organization needed an office large enough to house all its employees, the head office was situated in Dorval, a suburb of Montreal. While Denis and I were living in Saint-Eustache, I had been employed for a year or so at the head office. Living in the South, I was happy to work at something that was related to my own communities of Nunavik. I did administrative work for Tommy Gordon (also from Kuujjuaq), the head of the adult education department. When we moved back north to Kuujjuaq, however, I began training to be a student counselor at the high school level.

While at the Dorval office, I had learned about the board's partnership with McGill University in Montreal to offer training for would-be Inuit teachers, under the leadership of Doris Winkler. Using this type of model, Jack Cram and Andy Hum of McGill developed a student counselor program. Our on-the-job training courses were held at the KSB head office in Dorval and in some of the communities in Nunavik, on an alternating basis. Shortly after Denis and I returned to Kuujjuaq, I started to take these courses in counseling and education and to shift my work focus from health to education issues.

I started my full-time work in Kuujjuaq counseling students at Jaanimarik School (named after Lizzie's father, our long-time school janitor in Kuujjuaq). As a high school guidance counselor, I provided both academic and emotional support. Many teachers would send me students who were having problems with either absenteeism or behavior, such as acting out or not doing homework. I often found myself sympathizing with the students as I learned how tough it was for them to focus on school, given what was going on at home or in the community. They would also often tell me that their classes weren't engaging them. Of course, a big part of my job was helping students determine what goals they wanted to pursue after high school. I also spent some of my time replacing the often-absent administrator if the principal called upon me to do so.

Working in the school, with the young people from our community, made it impossible for me to miss how the wounding of the previous generation was having a dire impact on the next generation. If I had noticed changes in Kuujjuaq in the years following my return at eighteen, the social landscape had shifted once again in the few short years I had been living in Mont-Joli and Dorval. When I was first married, most of us in Kuujjuaq were still leaving our doors unlocked and the keys in our snowmobiles. But now Denis and I were hearing more and more often about home invasions: young men breaking into houses to steal booze or articles that they could sell to purchase alcohol or drugs. Public drunkenness and violence continued unabated. A sense of danger was becoming evident all around us. I remembered the evenings of my young childhood, playing outside as the sun set. Now this seemed unimaginable. With the number of vehicles circling town, along with people frequently partying outside, especially in the summer, young families like us had to be much more careful with our children. The safe, nurturing Inuit world of my childhood, a world in which family and community members were strong in their resolve, with a great deal of integrity and dignity, was in large part gone.

Each day at school, I could see how this social breakdown was having an effect on the daily lives of the kids in our community. Many children whose parents drank were struggling to attend school because of lack of sleep or nourishment. Some were now witnessing violence in their homes, and it was clearly affecting how the children felt about themselves and life as a whole. The teenage light-heartedness and excitement that I remembered from my younger years at home and in the Churchill days were rarely in evidence. The sense of optimism that we'd had was missing. Many students were performing below their academic potential or showing signs of drug and alcohol abuse. Sadness and confusion were setting in for many teenagers, but often they were unable to even name these feelings as such. Acting out in aggression was frequently the only way they had to express their deep-rooted emotions. And suicide, an act that had sent shock waves and disbelief through Kuujjuaq when that first young woman took her life, was now, tragically, something that our communities were experiencing repeatedly. I tried to remain positive in my work, but the new reality was disheartening.

AFTER SEVERAL YEARS IN KUUJJUAQ, Denis and I had become incredibly busy with the kids, our jobs, and a variety of community projects. Denis was working six days a week as well as overtime on calls for the airport and for house fires. He created a training program in house-fire response for the municipality of Kuujjuaq, and was also involved in ensuring the outdoor community skating rink was well watered so the kids could skate and play hockey.

Denis was ready for more manageable work hours and wanted to transfer back to Montreal. I struggled with the idea. I was also unhappy and was channeling some of that frustration into socializing. I didn't know if I wanted to stay in my marriage and leave my community behind. But I had made a commitment to keep my family together, so we all moved south again.

By that time our kids were going into grades three and four. We enrolled Sylvia and Eric in French elementary school and settled back into life on the West Island of Montreal. But this new phase of our family life came with its own challenges. Not long after the kids started school, Eric began to experience problems. During teacher–parent meetings, Eric's teachers complained that he was inattentive and not doing his work. And his report cards were uneven. This was not the bright, energetic boy I knew, the boy who had learned his alphabet before he could even talk in full sentences. So I had him tested and diagnosed, fighting for support from the schools and the teachers. The child psychologist who tested him told me that Eric had a high IQ, not surprising as learning disabilities are not usually connected to a lack of intelligence. If we could just get Eric through high school, therefore, he would fly. The words seemed appropriate—because Eric adored planes and flying. But they were daunting, too. *If we could just get him through high school.* It turned out to be a struggle, and a further education for me about how difficult it can be to help children succeed in school if there are any barriers at all.

I transferred back to the KSB head office. This time, I took a position as a student counselor for Inuit students who had come south to attend CEGEP (the Quebec college system that precedes university) after high school. I saw, once again, the immense difficulties that our students were facing.

Many of our students were struggling with lower academic skills than their southern counterparts. Some had not been well prepared to write essays or long answers to questions, and many had no practice with the critical thinking needed to deal with the more rigorous and demanding academic expectations and course loads. Some were also struggling with family problems back home; more than a few had already learned to use substances to cope with situations arising in our communities. All these challenges were compounded by living far from home. The students were scattered throughout the city, many living with families who knew next to nothing about Inuit culture or had little sensitivity to the challenges our youth faced. Many were homesick and feeling the loss of community. Half of the students were heading back home by Christmas. Having been a student away from home for eight years, I knew how difficult it could be living with families you didn't know. Other students had been placed in apartments in poor, often rough neighborhoods—environments that were not supportive of scholarly success. All these young people were having to adjust to a markedly different culture and to the hustle and bustle of city life. I felt for them, remembering my own difficult times in Blanche and in Ottawa, but I was also experiencing some of the same disorientation now that I was back in the South, getting used to city life again and to balancing full-time work with a young family.

Shortly after returning to Dorval, I became the department head for student services. During my time counseling in the South, I had become increasingly frustrated with how we were failing to prepare our students emotionally and academically for success in their post-secondary studies in Montreal. The student services department was supposed to provide ongoing, long-term support and assistance. Yet in reality, the staff were lurching from crisis to crisis—one day meeting a student who wanted to drop out and return home, the next rushing to take a student to a doctor's appointment or making sure the checks for the families the students were living with were sent out on time. Our office was more like a chaotic crisis center than an educational facility.

The staff did their best. Those who ran the department worked hard, were well intended, and cared about the students, but they could barely keep up with the crises that would arise. And at student services, for the most part, we had little direction and no concrete budget to work with.

In my first few months as coordinator, I tried to assess what could be done to allow the students to fulfill their personal and educational potential. I wanted to build a stronger support structure in Montreal that would help them succeed in their chosen studies.

Under my leadership, the Kativik School Board created a supervised residence for the first-year students, so incoming students wouldn't be dispersed throughout the city. Instead, they would live together under one roof with good supervisors, good tutoring, and good food. These residences would ensure that the students had someone keeping a closer watch over their academic performance and their social life. Furthermore, as coordinator of student services, I tried to ensure that they had access to a better living allowance and additional tutoring. Our students were financially supported through a KSB sponsorship program that was funded by the provincial and federal governments, under a grant for Aboriginal post-secondary students within Canada. Our team in student services, which included John McMahon, Anna Campagna, and Wayne McElroy (all former teachers who taught in our communities), felt that the students were getting insufficient funds for all that they needed to live in the city, so our team worked hard to ensure the budgets were increased.

Our department also recommended an improved screening process to better assess students' academic abilities before they arrived in Montreal. Simultaneously, we pushed for changes in our community high schools, so that students could come south with stronger educational preparation that would allow them to achieve their desired grades in CEGEP. I also increased the technical and counseling staff within the department in Dorval, hiring counselors who were formerly teachers in our communities, including Gail Richardson, who replaced John McMahon when he left for a job outside of the country. Together, we worked hard for changes.

As student services coordinator, I also wanted to address the problem of alcohol and drug addiction in our student population. I didn't want graduating high school students to have their futures compromised as a result of their addictions. What's more, I recognized that for young people, it's difficult enough to keep safe in the city. If they have an underlying addiction, students are at greater risk in terms of safety and health but also in their academic success. With these considerations, our team got funding from the National Native Alcohol Drug Abuse Program (NNADAP), based in Ottawa, to launch a drug and rehabilitation treatment center. We named it Isuarsivik, meaning "the place where one gains well-being." It would be housed in the first-year residence, as we felt a center under the control of our regular counseling department in student services, with leadership and staff who understood Inuit experience and communities, was more likely to be successful for our students.

The operation of the treatment center became far more challenging than I had imagined. It was difficult to find strong and supportive staff. Those coming to drug counseling work with their own history of substance abuse can have huge advantages over counselors who haven't battled with their own addictions. These rehabilitated men and women have intimate knowledge of the road to addiction and the difficult path to sobriety. They can understand their clients in a profound way. Perhaps just as important, their recovery experiences are powerful models for others to follow—they earn attention and respect because they've been there. But unfortunately, some of these counselors end up being "wounded healers." Wounded healers are people who try to help others by working in the field of addictions and rehabilitation, but do so before they have resolved their own issues. Wounded healers will usually bring the client, student, or patient only to the point they have reached themselves. This problem made it very difficult to find the right staff to deal with and support our students in the area of addiction. A number of the counselors we hired were too troubled—unresolved souls, as it were—to work effectively with students and other staff. In all likelihood, my own staff and I were also struggling with our own unresolved issues and bringing some of these limitations to our work.

Unfortunately, despite all the focus and commitment we gave to the endeavor, not to mention the blood, sweat, and tears, the treatment center became overwhelmingly dysfunctional. The issues we were dealing with were too large, too systemic for one team in one center to address. As my frustrations continued to grow and staffing problems escalated, we decided to shut down the rehabilitation program after only a year. It had been a real learning curve for all of us.

In hindsight, it had been extremely ambitious to try to provide treatment to young students newly arrived in a big city, where they didn't have their families or community to work with. My team and I had just taken on far too big a challenge to have any level of success. It was certainly one of those times I felt I was becoming overwhelmed just as quickly as my students were.

After closing the treatment center, we instituted a stronger screening process for addictions before the students could be accepted to the Montreal program. If the students admitted to us that they were struggling with addictions, but truly wanted to study further at the post-secondary level, they would have to complete a program at a drug and alcohol rehabilitation center before reapplying.

Once we removed ourselves from administering the treatment program, some stability returned to student services.

The Kativik School Board treatment program did eventually reopen in Kuujjuaq, with the funds we had secured for the program there. The treatment center still runs in my hometown under its original name, Isuarsivik.

AT TIMES IT SEEMED that the challenges for our students and for the student services department existed on every front. The college providing the post-secondary education for our students, for example, seemed badly out of touch. The Dawson College coordinator of the "Inuit Program" had made a trip north to visit our schools and meet with teachers and students in the late eighties. During that visit, he made a number of highly negative comments about the capabilities of our students and about Inuit ability to adapt to change. What's more, in 1988 he wrote *A Report on the Program Developed by Dawson College for Inuit Students*. From the report, we learned that our students' attendance at the college had been considered an "experiment," and that the coordinator, without consultation of KSB student services, had been assessing how our students were doing. And yet the college had provided little in the way of specialized curriculum, instruction, or support for our students. In fact, most of the college seemed oblivious to our presence and to the needs of our community. In a class on Inuit and First Nations history, which I spoke to as a guest, the instructor and the other students had no idea that some of their classmates were Inuit. I wrote a letter to the coordinator outlining the unfairness and lack of transparency in how our students were being treated and being judged.

Meanwhile, I felt growing frustration with our own role in our education system, and how we dealt with our students and their academic, emotional, and social challenges. As coordinator of student services, I was a member of the management committee of the Kativik School Board. The committee met weekly to discuss the issues at hand, whether they were administrative, management, or schooling related. At these meetings, I would raise my concerns about the inadequacies of our school system and attempt to talk about the issues, as I saw them. Through this persistent prodding, I was able to initiate discussions within the board on the emotional state of our children. Unfortunately, although in general I felt like I was part of the larger team at the KSB, I also felt that my views about how badly our students were struggling and how we might address that problem were not well understood by the group. I remained steadfast, however, in my opinions, as I felt the system was failing our students at every level.

While I knew that my critiques of the system created some tension between me and some of the school board upper management, the strain became even more evident during the mid-eighties, when another major Inuit institution, the Makivik Corporation, under the leadership of my brother Charlie, started a process to discuss and review the state of education in our community.

Both the Kativik School Board and the Makivik Corporation were created out of the November 1975 James Bay and Northern Quebec Agreement, the first major Inuit land claims settlement in Canada. This agreement had been negotiated with Inuit and Cree of Quebec, the federal and provincial government, and three Crown corporations—the James Bay Development Corporation, the James Bay Energy Corporation, and Hydro-Québec—that had begun a major hydroelectric project in northern Quebec. A regional Inuit organization, led by my brother Charlie Watt, Zebedee Nungak, and others, had been established as part of a larger movement of all Canadian Inuit to represent themselves in land claims negotiations with the federal and provincial governments.

This agreement addressed issues of health, social services, education, administration of justice, and other economic and social development policies, in addition to land ownership and financial settlements. After the signing, new institutions such as the Kativik Regional Government, the Kativik School Board, and the Makivik Corporation, as well as municipal governments and landholding corporations in each of the Inuit communities, were created.

It's important to note, when talking about the James Bay Agreement, that it was not unanimously supported by all the Inuit communities in Quebec, and much conflict and tension arose during the process. In fact, the communities of Puvirnituq (the birthplace of Zebedee) and Ivujivik refused to sign the agreement and for a number of years didn't recognize the authority of any of the organizations established through it. Even those who were instrumental in negotiating the deal recognized that in many ways it wasn't ideal. Mark R. Gordon, a childhood friend and second cousin of mine who was part of the negotiating team at the young age of nineteen, would later write,

> Eventually we accepted the James Bay Agreement. The Agreement was very important to my people because it allowed them to take control of their communities, especially the basic services. It allowed us to take control of our education and of our health services. Many of these services and benefits are what most Canadians take for granted as being their right to have. But the Native people of northern Quebec had to give up their land rights. They had to give up aboriginal rights to be able to

gain the basic services which southerners enjoy without having to give up anything. We knew it was not a perfect solution by any means.

He went on to make this analogy:

And as we slowly push and try to gain some of these things that we believe are rightfully ours, it reminds me of a story I heard about six people hunting only one generation ago. These six people were starving and trying to find food for their family. They saw a snowy-owl who had just eaten a lemming. Snowy-owls are very picky about their food, so they won't eat the insides of the animal but only the meat. These six hunters had to divide what the snowy-owl would not eat. Our trying to get legal concepts and legal rights recognized by the government is often like the snowy-owl—we often have to eat what he won't eat, and we have to make do with that. But hopefully that will give us enough energy to go on with the hunt.

Zebedee Nungak described the thinking behind accepting the agreement in his usual distinct manner:

I knew that it does not satisfy everybody's ideas of how Inuit in this territory should govern themselves. But it was also negotiated at a time when the political leadership in Quebec and in Canada was not very enlightened about how aboriginal rights ought to be recognized and what place they ought to be given in the political structure of the country. So in a way, it was a major accomplishment just to have reached an Agreement, incomplete as it was.... Yes, we understand we are not getting absolutely everything we wanted originally when we started the negotiations. But all of us regarded it as a step in the right direction that will launch Inuit in the villages and the region [toward] an ability that we never had before, to run our own institutions, to try to improve living conditions, and to have more control, have the control we never had when the federal and provincial governments were running their services.... It caused a fundamental shift in the power relationship between us as Inuit, former governees, and them as governments, who did whatever they damn well pleased, without any regard to how people in the territory saw themselves fitting in to the structure.

Some of the new institutions that Zebedee referred to were, of course, the Kativik Regional Government, the Kativik School Board, and the Makivik Corporation, as well as municipal governments and landholding corporations in each of the Inuit communities. In June 1977, the Makivik Corporation assumed the responsibilities of the Northern Quebec Inuit Association and was given the task of administering the compensation funds received from the agreement to generate revenues for economic and social development. Shortly after that, in July 1978, the official transfer of both the federal and the provincial school systems to the Kativik School Board took place.

The KSB, which was responsible for elementary, secondary, and adult education, became the first Inuit-controlled school board in Canada, and one of the first Indigenous school districts in the world. Much of the early years of the board were spent on the task of building new schools and extending services to all the communities. These operations took up enormous amounts of time and attention. Commissioners elected from each community and an executive committee worked hard to get the system up and running, and much of what the board accomplished was impressive. In fact, the KSB had, in a short period of time, established itself as a reputable Native-controlled education system that could serve as a model for others.

But there were also problems with the school system. As the years passed, it was becoming more and more evident that the system didn't have enough resources or the capability to design, develop, and produce strong programs for its students. It ended up providing a watered-down version of the Quebec curriculum, which failed to engage the students. The curriculum also didn't appear to prepare our young people well for further post-secondary studies once they left our communities. What's more, knowing that the key to success for northern Quebec or Nunavik was the establishment of future self-government structures, many people were worried that an under-educated population wouldn't produce the kind of leaders these organizations needed. Some also wondered if the disconnect they observed between the board and the communities was a result of the fact that the head office remained in Dorval, despite the initial plans to have it located in the North.

Parents and community members expressed many of these concerns, as well as requests for a more open review process of the board, through their regional organizations and through the Makivik Corporation, whose mandate overlapped somewhat with the school board's.

Specifically, the Makivik charter stated that part of its mission is "to relieve poverty and to promote the welfare and the advancement of education of the Inuit; to develop and improve the Inuit communities and to improve their means of action; and to foster, promote, protect and assist in preserving the Inuit way of life, values, and traditions." Seven years after the signing of the James Bay Agreement and the creation of the KSB, hard questions remained. Why were our children not succeeding in school? Why were so many dropping out? And why were those who did graduate from high school not making it in post-secondary? Why were the suicide rates still high and escalating?

During that time, I prepared a paper that I felt would help our organizations understand what was happening to our post-secondary students. In the report, I noted that the majority of our secondary-school graduates expressed the desire to pursue higher studies. Statistics, however, showed that the dropout rate from post-secondary studies was approximately 43 percent. One of the main reasons for withdrawal was academic weakness; many students had relayed that they did not feel they were challenged enough in high school and therefore were unprepared for post-secondary in the South. Involvement with alcohol and drugs and the struggle to adapt to a southern environment also played roles in the high dropout rate. Although there were many other reasons, including the absence of clear goals, relationship problems, family obligations, homesickness, or a lack of study skills, our experience with the post-secondary students in Montreal clearly showed that these factors didn't exist in isolation. It was a combination of these issues that led our students to withdraw.

I also noted that we were continuing to evaluate these main areas of concern in an attempt to improve the odds for our students. We had been trying to address the questions raised in previous forums, looking at how we could better prepare our students academically and improve the quality of our schools. This included conducting more evaluations and examining how transitions between levels of school, the performance of students, and social problems affected the ability of our students to learn. In addition, we looked at how resources were distributed within the board and between the schools, and asked tough questions about how we as a school board were evaluating new teachers who were sent to the North.

It became clear, however, as I worked on this report, that my supervisors and several peers were responding defensively and misinterpreting what I was trying to convey. They seemed to feel that I was attacking the system as a whole. One of my fellow Inuk coordinators accused me of minimizing the commendable work of Inuktitut teachers and Inuit pedagogical counselors in developing Inuktitut materials. These misunderstandings caused me much pain, as this was certainly not my intention. As an Inuk, I was proud of the work that had gone into developing those materials. But I was looking at the larger picture and trying to convey that Inuktitut materials alone did not necessarily make for a solid program. Nor did good ad hoc programs here and there, in either language, make for a strong curriculum, especially if a curriculum framework itself, the very foundation of any education, was missing. But my colleagues continued to view my concerns as an attack on my own culture and language.

I began to question whether I could remain with the Kativik School Board. Each time I had raised important struggles and issues facing our students, my comments had been met with defensiveness, rather than being seen as constructive feedback that would better prepare our students for success, academically, emotionally, or otherwise. And it was becoming clear that my voice was not welcome within the school board management and, in fact, was interfering with the majority voice of the managers. I wasn't willing to be muzzled in this way, and I declined to attend the upcoming Makivik education workshop altogether.

It had been many years since the signing of the James Bay and Northern Quebec Agreement. We had gained control of many of our organizations. We had embraced the education system we inherited with the same patience we showed out on the land, thinking that it would get better over time. In fact, that remarkable patience blithely added to the problems we were facing now: things were not getting better. Instead, those responsible for improving the situation had become distracted by politics and petty rivalries. It was our job to ensure that our kids got the education they needed to succeed when they graduated from high school. Not only that, but now that we were administering our own school board, it was our job to ensure our students were better prepared than they would have been if the federal or Quebec government had been running the schools. There's no point in gaining autonomy if you're not going to improve things, particularly the prospects of the next generation. But if that was the criterion we were to be judged by, we were failing, as a large majority of students didn't complete their post-secondary studies. The dropout rate was high, and many would be home by Christmas.

I thought this was a travesty, and I'm sure my colleagues on the board thought so too. And yet somehow we ended up not as allies but as opponents. Naturally, I was hurt by this, but I was also troubled in other ways. It was deeply discouraging to watch as we squandered a historic opportunity. These struggles were indicative of how quickly even our own organizations had adopted the worst aspects of institutional and bureaucratic culture. We were creating an "us versus them" divide within our community, just as our southern colonizers had done to us. We just couldn't see it.

In a memo to the administration, I told them that I did not want my Inuk voice to be silenced any longer. From the time I was sent away by the government for schooling, my voice had been controlled. I explained that after so many years at the Kativik School Board, I was going to take some leave to fully and objectively evaluate whether I wished to remain an employee of the board.

During my leave, I reflected on what I needed to do. My primary conclusion was that I needed to find ways to value my own voice. It was obvious that I couldn't do this working for the school board. Shortly after taking my leave, I resigned. I left in my own quiet way, without too much conflict, and without writing a detailed resignation letter.

As adults, we go through periods where certain childhood memories come back to haunt us and test us. Since the age of ten, when the Rosses had opened the letters I'd written to my family, I had struggled with the fear of retribution or censorship whenever I attempted to express myself honestly. This is why it took me a long time to sit down and write my detailed resignation letter to the Kativik School Board. And it was why I didn't present it to the board at my departure. It was only many months later, when the KSB opposed my involvement with the Nunavik Education Task Force, an independent review body, that I mustered the courage to share, in writing, my reasons for resigning. Ironically, my frustrating and painful experiences with the school board had pushed me to overcome my first wounding—the silencing of my voice in Blanche.

4

FINDING OUR VOICE

WHEN I WAS A CHILD, I had a number. Mine was E8-352. It was stamped on a small red tag, and on the other side was an image of an intricate crown. My mother had one too, as did the rest of my family and everyone else in the community. When I left for Churchill, my mother gave me this disc to take with me. In my youth, the little tag was a curiosity, without much meaning for me. I would later discover, however, that these small bits of pressed fiber or leather and the numbers stamped on them were part of the story of how our Inuit communities had been controlled and made to fit into a southern governing structure.

Before the trading companies, the missionaries, and the federal government arrived in the Canadian Arctic, our Inuit population did not use last names. This made keeping track of our population a challenge for the government and administrators. In the 1940s, to solve this problem, Canada began assigning each Inuk man, woman, and child a number. If you lived east of Gjoa Haven, your number started with E, and if you lived west of that point, with W. The one or two numbers immediately following this letter indicated what community you were from. The last set of digits was *your* number. By the time I was a teenager, most Inuit had adopted last names (as a result of "Project Surname," a program inspired by Simonie Michaels, the first elected Inuk of the Northwest Territories Council, and run by fellow council member Abe Okpik), and the discs and numbers were discontinued.

But the symbolism of these dog tags remains powerful. The discs remind us of a time when our Inuit communities were being forced off the land, relocated against our will, sent off to southern schools, and in so many other ways coerced into fitting into a social and political system that was not our own. They are a reminder of a time when *ilira*, or fear, made it difficult to speak for ourselves, to find our own voices.

But with the formation of the various Inuit corporations, as well as other Inuit international organizations, we would begin to take back control of our communities and speak for ourselves in the North and on the international stage. In the years that followed my time with the Kativik School Board, I became part of this powerful movement.

After my final difficult days at the board, I started to take on consulting work and contracts within my own region of Nunavik. In the fall and winter of 1990, I carried out a review called "The Validation of Health Objectives in the Kativik Region" for the regional health board. The focus of the review was on alcohol and substance abuse, and it involved traveling into several of our communities to interview health-care workers. I am not an academic, and in those years the statistical data were poorly documented, so the report was brief, but it did address the issue of addiction and the resulting consequences for the individual and for family life. Domestic violence, rape, incest, suicide, low family income, and teenage delinquency were at the top of the list. Alcohol and drugs were also a major factor in accidents, heart and liver problems, family psycho-social problems, poor nutrition, child neglect, and wife battery. The report drew further attention to the fact that the younger generation was experiencing secondary but just as serious effects. Having grown up within dysfunctional families and environments, children themselves turned to alcohol and drugs as a way to cope. These newly acquired habits took an additional emotional toll that included anxiety, insecurity, confusion, impulsivity, relationship difficulties, and suicidal tendencies. The report also addressed the socio-economic impacts of addictions on Inuit-run political, educational, and service-oriented organizations. Our institutions suffered from high turnover, absenteeism, loss of productivity, poor performance, lack of initiative, and the need for more staff as workloads increased with the rise in problems. In a short time frame, the link between alcohol and drug problems and crime rates also became evident, with increasing incidents of vandalism of buildings and property, breaking and entering, and robbery. In fact, the majority of cases tried and resulting in jail sentences in the South were alcohol or drug related.

This visit into the communities confirmed what I, and many others, already knew. The Nunavik I had grown up in was no longer. In one lifetime, we had come to be highly dependent on institutions, but also on alcohol and drugs, as a means to cope with the tumultuous changes to our Inuit world. In the report, I stated that the problem of addictions couldn't be dealt with by focusing only on the alcohol and drugs. The larger issues of how we lost our way, especially the loss of control over our lives and destiny and the loss of life skills, needed to be addressed. If we wanted to start the process of change from dependence to a healthy life, the life that people of my generation could still remember, we had to look at the lack of freedom of choice. I offered recommendations with strategies that included life-skills training in school and education about the psychological and physiological basis of substance abuse. In other words, I recommended that we give our youth some tools to avoid the trap of addiction. I also felt that the families and community needed these skills just as much as the youth and recommended that these types of programs be made available for all. I added that politicians and administrators of our organizations, not just the helpers on the ground, needed to learn about addiction so they could make appropriate decisions about funding addiction prevention and treatment. It was important for our health, education, and social institutions to search for and implement innovative, proven programs and adapt them to our situation, rather than offer programs that enabled or embraced the problems and allowed them to grow even more. I ended the report by recommending that alternatives with the potential to change the situation in our communities be sought, such as sports, nutrition, and self-improvement programs that would foster better understanding, freedom, and growth. Last, but certainly not least, I pointed to the need for more Inuit history and traditional-skills programs.

During the months I was busy working on this report, the Nunavik Education Task Force was being established by the Makivik Corporation. I shared the feeling of many leaders and community members that such a task force was long overdue, and I observed from a distance as the work of setting it up unfolded. I was interested in what the goals and objectives would be, and how the task force would be structured. So I was honored to hear that my name had been brought forward as a possible member of the group, although I had no desire at that time to have an official role in the work. What was less pleasant to learn was that the Kativik School Board had adamantly refused to include me. The board members argued that I wouldn't be an objective advisor to the task force, particularly in regard to the KSB, and they even stated this in an official resolution signed by their president and commissioners. Although perhaps not surprising, these actions by my former colleagues had a stinging effect on my spirit. It has taken me a long time to understand how my fellow educators, many of whom were Inuit, and all of whom ultimately wanted the same things I did, could have ostracized me in this way once I left the institution. But at the time, their actions left me feeling deeply hurt and desperately isolated. On top of that, I was having trouble in my marriage. In this dark period, I had started to party excessively on the weekends and to smoke more and more, in an attempt to alleviate my own pain. It was perhaps the first and only time I thought that life might be too difficult to carry on with.

As I was lying in bed one morning, struggling with these dark feelings, I got a call from Johnny Adams.

Johnny was among the most respected leaders in our community. He had been the mayor of Kuujjuaq for many years, besides having his commercial pilot's licence and running a successful hunting-and-fishing outfitting business for tourists. A friendly, approachable, and confident man, Johnny always reminded me of his late mother, who was one of the most gentle and kind women I recalled from my childhood in Kuujjuaq. But I will always remember Johnny himself as the homesick little boy whose tears I dried in the hospital in Roberval all those years ago. The connections we have as children have a long-lasting effect.

Johnny had been named as one of the Education Review Committee members of the task force. He told me he was calling because he had heard about the KSB opposing my involvement on the task force, but he felt that, with my ten years' experience with the board, I could be helpful. He also knew that I was passionate about the issue of education. I had never intended to be involved in this process, nor had I initially desired to be a part of it. After all I had been through with the board, I suppose I wasn't looking for any more arguments. I was battled out, in a sense. To have been spurned by my own, to feel that my colleagues and fellow Inuit had interpreted my views as something totally different from my intentions, had been deeply troubling and disheartening. But the flip side was the encouragement I got from people like Johnny. The closeness of our community can really keep you going when those around you lend you their strength. The fact that Johnny was to be on the task force as well, and that I was being invited by someone I respected and who shared my priorities, was enough to convince me to become the Inuk advisor to their work, along with the *qallunaak* advisor Bill Kemp and Wendy Ellis, who had been hired to help administer the work. I would be working for Minnie Grey, Mary Simon, Annie Tulugak, Josepi Padlayat, and Jobie Epoo, who made up the Education Review Committee (although Mary Simon and Annie Tulugak eventually withdrew, and Bill Kemp left as well, feeling torn between the two organizations).

Those of us on the task force would need that closeness and strength of community in the months to come. For some reason, misfortune swirled around all of us. It started with Minnie Grey's son, Aloupa, whose father was my cousin Willie. Minnie and I had traveled from our home base of Montreal to Kuujjuarapik, where the task force was presenting one of our initial reports to the Makivik annual general meeting. Word came that Aloupa had been rushed to the hospital with heart failure and urgently required a pacemaker. He was only sixteen years old. That was just the beginning of things to come.

That same year, Josepi was caught in a hotel fire in the dead of winter in Kuujjuarapik. He'd been attending meetings there with others from Inuit communities across Nunavik. He survived, and even ended up saving people, running outside in frigid Arctic weather barely clothed and in bare feet, but we did lose Paul Alaku from Salluit in that fire.

The third person to be affected was Johnny. He lost one of his tour clients in an unfortunate accident on a remote lake. Then, tragically, Jobie's son was killed in a four-wheeler crash. And I certainly didn't escape unscathed.

First, after seventeen years, my marriage broke down. My husband and I had met, married, and had our children before the age of twenty-three, so most of our young adulthood and early years together had been focused on making a home for our family, whether in Kuujjuaq or Montreal. But we were markedly different in our approaches to child raising, and in how we dealt with life issues in general. I had thought of leaving the marriage a couple of times over the years, but now I finally made the decision to end it once and for all. It was a huge personal challenge, and in going through with the split, I finally understood why the first couple of attempts had been so hard. First, to hurt another human being is always difficult. Telling Denis was tough. But there was an even more daunting hurdle to face. Once I made my husband aware of my decision, there sat my mother to confront. My mother, who thought the world of Denis, couldn't possibly understand or accept that I would leave him. A woman whose *qallunaat* partners had all left her to fend for herself as a single mother wouldn't take this lightly.

My sister, who had always been the buffer between my mother and me, forewarned her of my call to come, trying to soften the blow. I still recall the terror I felt as I was about to hear my mother's words. As I feared, she said she would never forgive me. How could I leave this good man when he was not leaving me? As difficult as it was telling my mother, my conflict with the school board helped to prepare me for this tough personal choice and the consequences that followed. Despite my mother's objections, I didn't change my mind. And as the years passed, my mother forgave me. She didn't exactly put this forgiveness into words, but I could see her pride in the passion I gave my work. She didn't realize, however, that it was her own ability to survive and thrive as a single woman that was my model and that carried me forward personally and professionally. I'm grateful to her for this. Denis also told me that he found help accepting our separation and subsequent divorce by recognizing that he was losing me not to someone else but to Nunavik. It didn't take him long to find a new partner, so life does have a way of working itself out.

After dealing with these challenges, what is surely every mother's worst nightmare came true. My daughter was hit by a car. She was riding a scooter outside her high school in Pointe-Claire, Quebec. Though her life wasn't in danger, her femur was literally shattered—pieces of bone protruded right out of her thigh. She was in the hospital for many weeks and underwent major surgery. Her doctor, Dr. Rhodes, told us later that he had initially thought he might have to amputate her leg, but he was a skilled surgeon, able to work miracles, putting the pieces of bone back together like a jigsaw puzzle. Sylvia had dreamed of becoming a dancer. Now, at the young age of fifteen, the enormous task of rehabilitation lay before her. She spent over a year and a half healing. I remember taking her out for walks in her wheelchair for most of that summer. She eventually moved on to crutches, and later a cane.

There's a common saying among us Inuit, *Tamanna Anigutilarmijuq,* which means "This too shall pass." I took solace in this thought. And in the belief that these moments, when life seems to be breaking down, often signal that we are on the edge of a breakthrough in our lives.

Indeed, those crisis-filled months only strengthened my certainty that I could make a difference. I realized even in my darkest hours that I was hardly the only person in my community who knew despair. Even in my small group, it was everywhere. And I was aware from my time in the clinics and the workshops that its shadow hung especially over our young people—many of whom choose to end their struggles, rather than pass through them. As a result, they lose the lesson that we are all here for: that we gain a larger evolution of spirit through our trials. For me, it is important to pass on the life lessons I've learned to a generation who, as a result of tumultuous change and intergenerational trauma, have, in large part, missed out on the training that teaches us to trust the process of life.

I have thoroughly reflected on these events as I prepared to write this book. In hindsight, these personal struggles developed my ability to deal with opposing voices and future losses, and they also strengthened me as a woman. My mother and grandmother had been role models of remarkable survival through great difficulty, and I was grateful to have been raised by them, in spite of my feelings of fatherlessness. I have also realized that this challenging time groomed me for things to come, for future difficulties in leadership positions. It instilled in me the ability to persevere and work with conviction on the things I believed I needed to say and do. It would be the training ground for the next steps in my political life, where I would encounter even more resistance and require even more perseverance.

During these personal tragedies, the members of the Nunavik Education Task Force became a close-knit team, and even more so, a family. What's more, the hardships helped us bring focus and strength to our work: the politics that threatened to steer us off track were nothing compared to what we were experiencing personally. In fact, these challenges brought more clarity and a deeper perspective to the issues. We were, after all, trying to make a better education system for our children.

Yet we were continually tested, not just by our individual challenges, but also by the escalating politics with the school board. The KSB insisted on having control over the project, even though it was clear this work was to be independent from both organizations that were providing the funding. They even wanted to veto our hiring decisions. I had invited Ken Low of the Action Studies Institute of Calgary to join the task force after hearing him give a talk on adaptive learning. I felt that Ken's research and work in this area made him the ideal person to put on paper what we Inuit knew to ring true from our own upbringing and life experiences. Yet many on the board felt that Ken, a *qallunaaq*, without enough academic credentials per se, was not "qualified" to review our school system. Another *qallunaaq* advisor working for the board clearly felt that there was no need to bring in new blood. But the task force forged ahead, and Ken became our chief writer, putting into words the details of the issues as directed by the team, and bringing to light the larger issues and the ways that institutional learning can contribute to problems.

There was clearly a need for our task force. In meeting after meeting, we heard from parents, teachers, students, and community members who were troubled by the shortcomings of our education system. Many people commented on the current curriculum, much of which was derived from or taken wholesale from southern systems. They noted that it wasn't appealing to and had no meaning for our students, making it difficult for them to feel engaged with their classes. Much of the material was also focused on skills needed in a southern setting—our children weren't being taught the necessary skills to survive their own environment and to deal with the tremendous challenges and changes facing our communities.

Indeed, our language, culture, and identity, along with the role of parents and the school board's responsibilities, were major topics of discussion. Many wondered if teaching our language in the schools was slowing down the process of learning a second language. Others felt that learning our mother tongue first was a must in order for young people to know their identity. Still others felt that teaching three languages—Inuktitut, English, and French—was a lot for a young school board to deal with. Some wondered if our children would be adequately prepared for jobs that required English and French if they were taught only Inuktitut during the first several years of schooling. Many community members emphasized the need for formal education to lead to employment.

In turn, many educators in our communities felt that parents needed to understand the goals and purposes of formal education more clearly, and they encouraged increased parental involvement. Culture and language teachers were calling for more support from parents, elders, and community members, in order to help our young people learn our Inuit language, traditions, and cultural ways. Teachers who had been working in the North for many years remembered a time when they had been more involved in community life, with more after-school interactions with parents and students. Teachers also commented on the extra burden of having to develop their own materials, coming up with their own lesson ideas and spending many hours making copies for their classes.

Interestingly, we heard from many students who said that they wanted higher standards and higher expectations for their work in class and at home, so they would be better prepared for what lay ahead in post-secondary education and as adults. A number of adults compared their own schooling, which had been delivered by southern school boards, to that provided by the KSB, noting that the former was more disciplined and rigorous. Many talked about their positive experiences at the Churchill Vocational Centre in Manitoba.

Overwhelmingly, what we heard, however, was that we had to define our own priorities in our schooling system based on the foundation of our language and culture. In particular, our children needed to learn more survival skills and kinship—both of which are essential aspects of Inuit culture.

When southern institutional education systems were set up in our communities, or when we were sent off to schools geared to Inuit children in another part of Canada, the new type of education reflected southern cultural understandings and ways of knowing. It did not fit into our culture or our language, nor did it address the challenges that we Inuit were facing, mainly the tumultuous changes that had brought new ways of life. The institutional approach to learning was thrust upon us and all but eliminated the traditional way in which we educated our children—through the discipline of the land. With yet another loss of control, we wondered how we would continue to maintain influence over our children's education.

Yet, for my generation, the southern education we received was positive in a number of ways. As I mentioned in describing my own time in Churchill, despite the many negative aspects of this "deprogramming"—the loss of our ability to speak our language, or to dress, act, eat, and even think in our Inuit way—the schools, for the most part, were demanding and instilled a sense of discipline. Although the content and subjects were highly irrelevant to our world, the structured environment in some regards reflected the disciplined, holistic way in which our Inuit culture educated children for the opportunities and challenges of life. Not only did many of us feel challenged, but also, as we progressed academically, at times surpassing our southern peers, we came to believe that we were equal in our learning ability to those in the South. Many my own age talked about these times during public meetings with the task force. But we had to acknowledge that the traditional foundation of our lives gave us the ability and the resilience to survive in a rapidly changing world. Sadly, the next generation lacked this foundation. They spent less time in nature and on the land, the natural teacher of life, and they suffered from the unresolved traumas of their parents and those they loved.

On top of that, over the years, the academic rigors and expectations of our education system had been lowered, due in part to a general North American trend to lower standards, and in part to our government's shift from a paternalistic to a maternal approach in teaching Aboriginal children. It had reached the point that many of our youth were no longer challenged. (Our communities still talk about "social passing"—moving students from grade to grade based on age rather than academic achievement—and how it must stop if we want to be realistic about our children's abilities and help prepare them for post-secondary education.) What's more, we observed that the current schooling system was entirely prescriptive, failing to give our students the control they needed over their own learning. Indeed, our students were hungering for challenge and were seeking it instead from other means, like video games or, worse, risk-taking behaviors. (In my opinion, video games in and of themselves aren't bad. Many unchallenged kids build skills such as strategic thinking, boldness under pressure, risk assessment, and critical analysis by playing video games. But developing these skills exclusively indoors, instead of out in nature with elders or parents as guides, can become problematic.) Our young people's thirst for challenge and control was being shunted into habits and acts that weren't necessarily survival based or that failed to offer the guidance that would help them make wise choices. This led them all too often to more self-destructive patterns.

The many discussions the task force had with our communities made one thing clear to all of us: if our education system did not respect or challenge the creative potential and intelligence of our children, it would crush them, not liberate them.

The issues surrounding our school system didn't exist in isolation, of course. In those years, and today, the loss of our children's autonomy was a reflection of the growth of many other dependency-producing institutions in our Aboriginal communities. As social problems increased over the years, programs and agencies were introduced by the federal, provincial, and municipal governments to "deal" with them. Yet these institutions seldom empowered our community, nor addressed the deep historical traumas that had created our wounds. Instead, financial support, counseling, and law enforcement often took authority away from our elders and communities. It should have come as no surprise that if young people didn't see their elders take command of their lives and their culture, they would quickly lose their own sense of responsibility and survival. Even institutions as seemingly benign as social and health services and the courts had been deeply corrosive to a culture encouraged to rely on them, when for millennia they had relied only on themselves for direction and acquiring wisdom. The task force recognized this.

The task force's observations circled back to the issues I had been looking at in my earlier health report. It was difficult to ignore just how many of our own people continued to rely on the thrills of gambling and sex, or on handouts such as welfare. Nor could one overlook the insidious loss of personal freedom and morality caused by addictions to alcohol and drugs. It was heartbreaking to see how easily we could become dependent on things we didn't need—indeed, things we had never known until recently. What we needed to understand was that the problem of dependency could not be solved through reliance on dependency-producing institutions. Rather, we had to find concrete ways to encourage people to trust themselves and those around them enough to free themselves of their reliance on others. Addiction, poverty, and food insecurity could not be separated from the larger issues of freedom, self-worth, self-reliance, and self-determination, and the solutions had to be holistic if we were ever to shake ourselves out of the paralysis that seemed to have gripped our communities.

Our future, in many ways, lay in what we knew from the past but had mislaid. In Inuit culture, and, in fact, in all Aboriginal cultures, our elders were the source of wisdom, holding a long-term view of the cycles and changes of life. Wisdom was forged through the independent judgment, initiative, and skill required to live on the land and ice. The land was our teacher, and our hunters knew the value of patience and trust. Without them, they would perish.

That's what the Nunavik Education Task Force saw and heard, and that's what we wrote.

The Nunavik Education Task Force put together a trilingual report in Inuktitut, English, and French entitled *Silatunirmut: The Pathway to Wisdom*, which outlined our 101 recommendations for change. It walked readers through the education system in the North, noting what had happened in the past, how it translated into the problems in the present, and what changes would allow for improvements in the future. (For a more detailed account of what was said during the task force meetings and in the final report, readers can turn to Ann Vick-Westgate's book *Nunavik: Inuit-Controlled Education in Arctic Quebec*, published in 2002.)

Our report argued that any effective education system must consider community needs, including self-government, cultural preservation, and development of community and regional infrastructure. We continued by saying it must also, as a means of empowering our children, seek to improve self-management skills, heritage and cultural survival skills, analytical skills, and community and economic skills. Our students also needed to acquire global cultural access—in other words, they needed to learn about the world beyond their own doorstep, including environmental, political, economic, and business issues—so they could see global changes coming their way and interact with the world more effectively. Global cultural access would be an important part of giving them a sense of control. By making ongoing changes and reforms with the required political commitment, leadership, and determination, we could empower ourselves and regain control over our lives and our communities.

In other words, to amend the education system, we recommended changing *everything*.

NATURALLY, OUR REPORT PROVOKED a backlash. It would have been naive not to expect it, and though it was hardly welcome, we were ready for it. Although I had been hired primarily to help guide the research and work with the main writer of the report, Ken Low, I was asked to stay on to start the process of implementing the task force's recommendations.

In the beginning, putting the report recommendations into practice was painfully slow work. Everything had to be approved by the commissioners of the Kativik School Board. Over the course of the two years I worked at the Makivik office, however, the process began to open up. A shift in leadership at the KSB, as well as an initiative called *satuigiarniq* ("reclaiming"), overseen by my cousin Annie Grenier, who had become director general, and executed by Sarah Bennett and Elaine Armstrong, created an environment of inclusion and acceptance, where the new path and new ideas could be embraced.

As fulfilling and important as the work at Makivik was, however, it was a failure I experienced at that time that may have inspired me the most. In 1993 I ran for the office of corporate secretary of Makivik. Although I never thought I would run for elected office, I had begun to realize that as hard as I worked over the years, at every turn I depended on the goodwill of a small number of peers who invited me to become involved. These were good people, whose goals I shared, but it dawned on me that I couldn't wait for the next great opportunity that happened to come my way. I couldn't sit still. If I wanted to bring about the change I increasingly felt needed to happen, I would have to go out and earn my mandate.

Of course, it's never that easy. When you're attempting to enter the political arena, you'll always face challenges. As a woman it is that much more daunting, trying to infiltrate the old boys' club. Our region was certainly no different from anywhere else. In the history of Makivik, only three women—Mary Simon, Minnie Grey, and Martha Kauki—had ever become an executive member of the corporation. From the outset of my campaign, I had plenty of opportunities to consider the way the cards were stacked against me.

Although the people of Nunavik had always known who I was, the controversy and tensions surrounding task force work had instilled doubts as to what I really stood for. In particular, my struggle with the Kativik School Board, made public by the board's official letters and resolutions objecting to my involvement in the task force, had called into question my integrity and honesty. Of course, elections are not meant to be friendly affairs. Even so, I was shocked that so much of the mud that had been thrown still stuck. I lost the election.

But there's nothing like a good public debate, where others question who you are, to force you to figure yourself out. While campaigning, you're constantly talking to people about who you are, what you believe in, what you stand for, and what you feel could improve the well-being of your communities, of your youth, of your school systems, of all your organizations. You're putting yourself out there in a very raw and vulnerable way. As the public starts to get to know who you are, you yourself come to understand the same thing at a deeper level. Sometimes you never fully know who you are or the stuff you're made of until you're forced to fight for what you think is most important.

In hindsight, that defeat was something I needed in order to get back on the path to the person I was meant to be. In the heat of the battle, my true self emerged, and I had never felt so powerfully connected to my voice and my own resolve to become a leader for our people. Moreover, I realized that the election *result* didn't determine whether this was a worthwhile task. Instead, it was the *process* that was building my capacity to participate in electoral politics.

That process cast me into the public eye more than I had ever been, and although I had the sense that I was doing something that needed to be done, the experience wasn't altogether comfortable for me. I don't suppose many people would guess that I am by nature an introvert, but I am. I've never been comfortable being a public person, despite the type of work I ended up doing. And I haven't always found it easy to speak in public. While I was in high school in Ottawa, in fact, I was so rattled during a class presentation that my mouth went dry and I could hardly get any words out. When I did eventually start to speak, I stammered through my entire talk. I think the teacher, feeling sorry for me, gave me a decent mark only because of the content of the paper itself. And certainly as a young adult, I would never willingly put myself into a position where I had to speak to a crowd. It was because of my commitment to help our communities deal with social issues that I projected myself into the public sphere in such a big way. Stepping up to the plate to campaign for political office with Makivik Corporation meant that I was moving well out of my comfort zone.

Not that I found it discouraging. Still living in Montreal, with my teenaged children, I ran again two years later, and threw myself into the campaign with even more vigor and conviction. By then the communities had come to realize that I was the same person I had always been, and I was elected as corporate secretary for Makivik Corporation in March 1995. I had run on the mandate of addressing the social and educational challenges of our youth in Nunavik. I'd always been a strong proponent of creating social and educational structures that would improve the lives of our youth. The political power base I gained through my election to this position allowed me to work on these important issues.

I also had the benefit of working with good, supportive people. Zebedee Nungak, the president of Makivik at the time, welcomed me in my new role, as did the other elected executive members. (In fact, long-time treasurer Pita Aatami—Johnny Adams's brother—would later become my greatest supporter throughout my international work as a board member of the Inuit Circumpolar Council. Pita and the executives at Makivik never wavered in their support, and this consistency helped me to be stronger in my stances internationally.)

During my years as corporate secretary, my children were becoming young adults. When Sylvia turned eighteen and graduated from high school, she moved back to Kuujjuaq to work. And Eric—his school years had been a terrible struggle for both of us. I had pulled him out of the French-language school after grade five, feeling that he would get more special-needs support in the English system. Things went much more smoothly for a number of years after that, and then he hit high school. I kept hearing the psychologist's words, "*If* you can get him through high school ..." But both he and I battled through it, and Eric graduated a year after Sylvia. He immediately entered a pilot-training program just outside of Montreal. Finally able to follow his passion, Eric thrived in the program, and before we knew it, he was a certified commercial pilot. Or he would have been if he hadn't been just seventeen—still too young to be officially licensed. The day he turned eighteen, he became certified and started his career with Air Inuit. During his training and early months on the job, he would come home on weekends, and he remained with me for the next couple of years.

My role as corporate secretary was to oversee the daily administration of the organization, including the preparation of all documents in Inuktitut and English for the executive and board meetings, and for the annual general assembly. But there was room to do other work as well. One of the first things I wanted to do in my new position was create something for the youth that would endure. I didn't want our funds to go into workshops or conferences, which often failed to lead to the kinds of changes that were discussed. Instead, Bob Mesher from Makivik; Gail Richardson, a student counselor for the Kativik School Board; and I created a film focused on youth and community development, called *Capturing Spirit: The Inuit Journey*. When we first premiered the film, it was well received by the leaders who attended the Makivik annual general meeting and by the young people we had written it for. In fact, the film continues to be shown in some schools and continues to air on the APTN TV network at least once or twice a year.

Writing, directing, and producing the film was certainly a huge learning experience for me, and it's a legacy that I'm proud of, but my three years as corporate secretary with the Makivik Corporation also provided a rapid education in the process of politics—both big picture and small.

During my entire youth and into my adult life, I had been a heavy smoker. When I reached the age of thirty-five, I quit. Later on, I developed an aversion to smoke and couldn't bear to be around people who were smoking. Unfortunately, most of the members of Makivik Corporation were heavy smokers, just as I had been. In the boardroom during our meetings, the air was grey and the table was cluttered with ashtrays. I could barely stand it.

Then I realized I didn't have to. I was in charge of corporate administration, and I set the agenda for the meetings. So my very first agenda included a suggestion that we create a smoke-free boardroom with "healthier air." The resistance to my suggestion was much more forceful than I'd expected—I remember mockery of my choice of words and angry body language. As a former smoker, I should have known I was picking a ferocious fight. However, we got over it, and I think people eventually appreciated the change.

But the small bureaucratic struggles I faced within the organization were nothing like the challenges I experienced being in the public eye. In my role with the corporation, the media's interview requests were at first minimal, but eventually they became more and more frequent. I was also constantly required to deal with the broader public. I recognized that this was necessary work in my position, but it was taxing for an introvert like me. I found the public interaction depleted my energy more often than not. But the difficulty I had being front and center, as well as all the other obstacles the job presented, made me realize that struggles follow you for a reason in life. They follow you until you learn how to overcome them, and come to clearly understand what your life's intention is: that the greatest challenge is to remain true to yourself and to your beliefs. The self-reflection and growth I went through in those early days at the Makivik Corporation were good things. My public position may have been a baptism by fire, but hotter fires were waiting right around the corner.

IN THE SUMMER OF 1995, only a few months after assuming the role of corporate secretary, I attended the Inuit Circumpolar Council general assembly in Nome, Alaska. Founded in 1977 by Alaskan Inupiat Eben Hopson, the ICC is an international non-governmental organization (NGO) that represents the

Inuit populations of Canada, Alaska, Greenland, and Chukotka, Russia—about 160,000 people in total. Its mandate is to protect and promote Inuit culture and way of life, as well as to represent Inuit interests and rights in cultural, political, and environmental concerns at the international level. It holds a general assembly every four years in one of the member countries.

In 1986 I'd attended one of these assemblies as an observer. This time I was a delegate representing the Nunavik region. Traveling to one of our Inuit communities in another country is always a great thrill. So much excitement fills the air as we gear up to discuss the issues affecting us at the international level. Many ICC players have been involved in this work in elected and supportive roles for years, so there are many familiar faces and well-known leaders from our Inuit world. Some have been with the ICC since the very beginning, which speaks to their commitment and sense of responsibility toward protecting the rights and interests of those of us who live on top of the world.

The assemblies are also a wonderful way to celebrate our Inuit culture. While some traditional dress may feature fur and others not, while our throat singing may differ from region to region (Greenland, for example, doesn't have a throat-singing tradition), and while our country food may be prepared in slightly different ways, the commonalities are powerful. During the events we all eat *muttaq*, fish, caribou, berries, and other Arctic delicacies together. The entertainment provided at night showcases the richness of our shared culture, with gifted and uplifting performers who drum dance and throat sing. Others give more contemporary performances, influenced by the Scottish and European whalers who introduced the accordion and fiddle to the Arctic peoples. Above all, the time together allows us to honor our shared traditions and culture as one.

Along with the elected Inuit leaders, many high-level politicians and representatives are invited from various international organizations, including the United Nations and, depending on the focus of the particular assembly, the World Bank, the World Trade Organization, the Convention on Biological Diversity, and the International Union for Conservation of Nature. Several top government officials from the four countries where we Inuit live are also invited, and some make brief comments to the assembly.

The assembly usually lasts four full days as Inuit of the world discuss their priorities and set out a work plan that will give the ICC leaders their marching orders for the next four years. At the end of these assemblies, I have always felt enriched by the connections made and invigorated by the renewing of the mandates and by the sense of family I get from working with fellow Inuit from all four countries.

At this particular assembly, what I heard resonated with me and touched me powerfully. At one session, chaired by Igmar Egede, a fellow Inuk from Greenland, I heard in detail about toxins that were being found in our country food. We were told that since the 1980s, scientists had been discovering that persistent organic pollutants (POPs) were poisoning the Arctic food chain, resulting in devastating effects in our populations. Igmar Egede had been working on this issue for a number of years, and he clearly and passionately explained what this meant for our communities in terms of our health, our environment, and our culture.

As I listened to Igmar, I saw that it was all connected. Our challenges were local, but they were part of something global. Our problems were not made in the Arctic. So our solutions couldn't be entirely local either. We could not look only inward.

Other health, education, and social challenges were fresh in my mind from the regional work I had been doing. I wanted to make sure that the global environmental issues affecting our Inuit world were not disconnected to the many other health-related challenges we were facing in our communities. I made it clear that I hoped that part of the mandate that the ICC was working on would be the search for solutions every bit as far-reaching as the problems.

My comments were well received, and I was well supported from the floor. I didn't realize how strongly supported I was until the nomination process for the new ICC executive began.

At each ICC assembly, an international chair for the executive council is elected, as well as eight council executives. The council executives are the chair and the executive council member of each country's ICC office. Elections for these offices are also held at the international general assembly. During the run-up to the nomination process, two other delegates from Canada, Eddie Dillon from the Inuvialuit Settlement Region and Gary Baikie from the Labrador region, came to me and said that they were going to nominate me for president of the ICC Canada branch. I responded that I didn't want the position, that I'd just been elected to the Makivik Corporation in March. I was still green, still learning, and I certainly couldn't take on a second mandate when I was so new to the political arena. They responded that they felt the busiest people usually got the job done, and that I could do this. They really put the pressure on, stating that they were going to nominate me regardless, as they thought I was the best person to head the Canadian branch of the ICC.

And that was that. Despite the fact that I'd had no intention of running for another leadership position, I was elected president of the Canadian branch of the ICC along with Joe Kunuk of Iqaluit. I still teasingly blame Gary and Eddie for starting me off on this international political life and all that has come with it.

My new role was to represent the interests and rights of Canadian Inuit people internationally. But the ICC organization, depending on budgets, will often allow one national branch to take the lead on a specific issue. Such was to be the case for the ICC's work on POPs. I became the spokesperson for all four member countries, along with Aboriginal peoples of northern Canada, including the Yukon First Nations, Dene, and Métis.

This marked the point at which my life's work found me. I was given an international mandate, and it dictated the direction I would now be taking. Having been given these two positions, I hit the ground running. I was propelled into the regional, national, and international world of politics.

During my public fight on education issues I had become very outspoken. But this strength came out of my sense of responsibility to give voice to the issues, no matter how taxing it was for me. I had to learn how to do it: in other words, it was a practiced way of living, not a natural one. But if I thought that I might be able to return to my introverted life, a comfortable place where I would safely do my low-profile regional work with the youth and the communities, the universe, it would seem, had other plans for me.

For the next three years, I held the positions of corporate secretary for Makivik Corporation, with its regional mandate, and president of ICC Canada, with its international focus. This kept me extremely busy. Not only was I dividing my attention between the two organizations but also I was immediately engaged in long-distance travel for the ICC, as well as frequent trips between my home in the West Island of Montreal and the ICC Canada headquarters in Ottawa. When I look back at those first three years, I often wonder how I survived. I do, however, give credit to the staff I had at the time, who were highly competent and supported me as I got to know the ropes of the two organizations.

My travels with the ICC underlined the paradoxes of our international Inuit community. We live in close proximity to one another, relatively speaking, yet we can be divided by great distances. The countries we live in may be strikingly different, yet our way of life, at its roots, is deeply familiar.

As Inuit, we are one people living at the top of the world. If you were to look at a globe from the top, rather than from the equator, you would see that northern Canada, Alaska, Greenland, and Siberia are all relatively close to one another. It looks as if we Inuit should be just a short hop away from one another by plane. But in truth, it can be a long process to get from one international Inuit community to another. The way in which airline flights are routed creates an extraordinary distance between us. For example, if you, as an Inuk living in Nunavik, wished to visit Chukotka, or, for that matter, Alaska, it would take you longer to get there than to Europe. If you lived in Iqaluit, Nunavut, however, and wanted to go to Greenland, it would take only an hour and a half to get to Nuuk, Greenland's largest city—if there was a direct flight scheduled. On the other hand, if no direct flight was scheduled (they are not always available), and you couldn't afford a chartered flight, you'd have to fly south to Montreal or Ottawa then fly through Copenhagen, or go farther down to Boston and take a flight to Iceland then over to Greenland.

At no time have the physical distances that divide us been so clear to me than during my first trip to Siberia, in 2003, as newly elected ICC chair. After chairing an ICC executive council meeting in Alaska, I flew from Nome to Chukotka with a fellow Inuk woman from Russia, who was now living in Ottawa and working for ICC Canada. Appa hadn't been home to see her family for several years. The family reunion at the airport was emotional. Appa was embraced by her mother and other family members. Her mother was elated to see her daughter and was grateful that I had chosen to bring her with me to spend a few days with her family. I still have the beautiful beaded slippers she made for me in gratitude for bringing her daughter home for those few days.

When we landed in Siberia, armed Russian men in uniform met our chartered plane. It felt like a movie scene. As we left the airport, I saw that the area was run-down and the cement buildings were in ruins. Appa's hometown, Provideniya, was nestled between two incredible mountains, but the area looked as if it had been through a war. The houses and buildings were almost all cement and steel, unlike in the Canadian Arctic, where our homes were made mainly of wood. Although it was a mere hour and a half away from Alaska by plane, I felt as if I had arrived in a different world.

Yet, despite the differences, there was still that connection to family, to fellow Inuit, that was just like coming home. We watched drum dancing and listened to throat singing. I was fed *muttaq*, and while another type of whale was used, it was nevertheless the same food. Though far from home, in an unfamiliar setting, I was home with our people. At no other time did I feel so clearly that we Inuit were all one.

The trip also gave me a window into what Siberians had been going through, both Inuit and non-Inuit. I stayed with a Russian woman who ran a bed and breakfast in her home on the second floor of a concrete apartment building. She told me that when the ruble (Russian currency) collapsed, the people of Siberia were all but forgotten. Her granddaughter, about five years old at the time, had come to live with her. The heating and electricity were shut off at certain times of day, so they would all go to bed wearing their clothes and their fur coats. The little girl would sleep between her grandfather and grandmother to stay warm. The woman also told me that her granddaughter wanted to color and write, but after a certain time of day, there was no electric light by which to do so. Instead, the young girl would work by candlelight. Straining to write in low light for weeks on end, the young girl's eyesight was partially destroyed. It was a touching story about the day-to-day struggles of our fellow Inuit and all Russians who lived in Siberia.

Members of the ICC had heard many such stories of suffering in Siberia after the collapse of the ruble in 1998, even tales of surgeries being conducted by candlelight and people eating dogs to survive. In response, in 1999 Canada's Arctic ambassador, Mary Simon, and ICC Canada had launched a humanitarian assistance program, dubbed the "Inuit Express." With funding from the Canadian International Development Agency (CIDA), the ICC flew boxes of food and supplies into Siberian communities, working alongside various agencies within the Canadian government, the Canadian Red Cross Society, the Canadian embassy in Moscow, and Russia's equivalent to the ICC, the Russian Association of Indigenous Peoples of the North (RAIPON). After the mission, *The Globe and Mail* reported that the program had been a boondoggle—that the bulk of the money had gone to the ICC's administration, charters, and airlines instead of to starving communities. Yet I remember families and ICC staff in Ottawa working feverishly to pack those boxes and get them onto the Inuit-owned First Air chartered plane. In the end, the

ICC was able to cut through a lot of bureaucracy, and some of those Siberian communities did get food and other staples that they simply wouldn't have received without the Inuit Express.

The ICC continued to work with our Russian counterparts during my term there. Indeed, my work with the ICC had me traveling to many international destinations for meetings and negotiations. I made several particularly memorable trips to Russia and South America for projects that ICC Canada had started before I was elected.

The main mandate of the ICC is to represent the interests and rights of Inuit people at the international level, working with the United Nations and other forums for those protections. However, we also had opportunities to reach out and share with other Indigenous peoples what we Inuit had learned over the last twenty years or so. Even though our Inuit organizations, institutions of learning, and social structures continued to face challenges, we had gained a wealth of experience over the years in negotiating our land claims, running our own businesses and airlines, and co-managing wildlife with the federal government. In negotiating our land claims agreements, we focused on our right to harvest wildlife throughout our settlement areas, as well as on compensation for development, should it affect the Inuit harvesting economy. We also created management boards that would work with the federal and provincial or territorial governments to manage our terrestrial and marine wildlife. Other boards were created to manage and plan for the development of land and water and to evaluate the impact of such development. National parks were established as a result of our efforts. We also created councils to help preserve and display Inuit archaeological heritage. In short, these land claims agreements were important pioneering work linking our economic, social, cultural, and environmental considerations in decision-making, and allowing Inuit to be part of the political development in our regions and to manage our institutions.

We saw that we could share a lot of these experiences with Indigenous peoples in other parts of the world who were just beginning to take back control of the management of their land and resources.

Before my time as an elected official with ICC Canada, Inuit leaders and representatives of the Maya and Garifuna peoples had connected at a conference on economic development for Indigenous peoples held in Belize. The plan was to create the Belize Indigenous Training Institute with the Indigenous peoples of Belize, guided in part by the Inuit experience of recent years. This project was already under way and was managed by Kevin Knight, a consultant for ICC Canada. We brought down a used transmitter all the way from Kuujjuaq, delivered by the mayor, Michael Gordon, and Craig Lingard, which provided some Belize communities with regional radio for the first time. Interestingly, the ICC struggled to find funds within our own country to work on national issues, as it was an internationally mandated NGO. But building capacity with fellow Indigenous peoples in Belize and in Russia—developing and delivering programs that would share the expertise we had developed through our land claims work—helped the ICC tap into much-needed funds from CIDA. This funding allowed us to also do our international work on the issues affecting our own Inuit homelands.

When we went into Belize for specific workshops, we invited members of our communities from Canada, Alaska, and Greenland who had expertise to share with others. I recall one particular workshop on co-management regimes where we brought together the Garifuna and the Maya, including the Kekchi Mayan people. It was the first time all three Indigenous peoples of Belize had been in the same meeting to discuss their issues and priorities. I recall one elder named Austin Flores who at the end of the meeting thanked us Inuit for bringing them together, saying, "My self-esteem went up a few notches today and I am grateful for this work we have started together." I was pleased to hear that the sharing culture of Inuit went beyond our borders.

Indeed, it was gratifying that while the ICC was in part financially supported (via CIDA) by projects like the one in Belize, we received and administered the money by helping others find their ground. The ICC, being a non-governmental organization, depended on funds from government, foundations, and other sources to fulfill its mandates. Getting this funding from CIDA was crucial, as it allowed us to carry on our mandates internationally, such as our global involvement with the POPs negotiations.

Our project with the Indigenous peoples of northern Russia was similar but more extensive, involving an Indigenous-to-Indigenous component as well as a government-to-government component. The Russian government wanted the work to include all its northern Indigenous peoples, not just Inuit, so we worked in partnership with the ICC counterpart in Russia, RAIPON. The Russian and Canadian governments were also involved in this capacity-building project, which was managed by Oleg Shakov and called the Institution Building for Northern Russian Indigenous Peoples' Project (INRIPP).

As with the Belize project, our co-operative work with Russia proved to be effective. Sergei Harushi, who was the president of RAIPON during my initial tenure with the ICC, stated publicly at an Arctic Council meeting, "We used to wait many weeks to receive responses to our communiqués through regular mail to our remote and northern communities in Siberia spanning many time zones. Since the capacity building began with ICC, we now worry when someone does not respond the same day to our emails." Our program helped the northern Indigenous peoples to establish regional offices equipped with computers and staff, and, on a larger scale, enabled them to be active with other circumpolar Indigenous organizations in the Arctic Council process, as well as in many UN programs affecting the communities of Indigenous peoples of the Far North. ICC Greenland also had a project with RAIPON, although smaller in scale than that of ICC Canada. We did what we could to help our less fortunate counterparts in northern Russia.

My work with the ICC Russian initiative took me to Moscow for meetings several times during my time as an elected official for ICC. Although dealing with the time zone difference, along with the pollution, proved to be taxing to my health, I feel privileged to have been part of this work. In the few short years that I was involved with ICC Canada, I saw Moscow and parts of Russia change from places of destitution to vibrant cities. In fact, I recall the first time I was in Moscow with Mary Simon, the circumpolar ambassador for Canada, and Walter Slipchenco, a veteran civil servant who spoke Russian and knew the lay of the land. It was early December, and my birthday came around while we were there. The three of us ended up celebrating at a Pizza Hut, the only place we could find to get a decent meal. Just a few years later, Moscow seemed like a different city, with stores and markets offering many more choices and the people appearing to have a considerably higher quality of life. I don't know how they managed to turn things around so quickly in their "southern" cities, but the speed of change did impress me.

The years I spent working for both the Makivik

Corporation and the ICC were certainly challenging. Challenging, but far from impossible. I came to realize that, fundamentally, I had a single mandate: to help improve the lives of our people in our communities. When I looked at the obstacles in front of me in those terms, no challenge seemed insurmountable. In fact, what I kept discovering was that to address the problems I might find around the corner, I had to go looking for solutions on a bigger and bigger scale. I had started out just trying to help the students who passed through my office, but with an international mandate, and under the auspices of this remarkable institution, the ICC, I, along with my ICC colleagues, ended up taking up the cause of the entire Arctic.

5

POPs and the Inuit Journey

DISTURBED BY THE INDISCRIMINATE USE of synthetic chemical pesticides after the Second World War, aquatic biologist Rachel Carson reluctantly turned her focus from nature writing to warning the public about the long-term effects of misusing pesticides. In *Silent Spring* (1962), she challenged the practices of agricultural scientists and the government and called for a change in the way humankind viewed the natural world.

Carson was attacked by the chemical industry and some in government as alarmist, but she continued to courageously speak out, stressing that we are a vulnerable part of the natural world, subject to the same damage as the rest of the ecosystem. Testifying before Congress in 1963, Carson called for new policies to protect human health and the environment. Rachel Carson died in 1964 after a long battle with breast cancer. A witness for the beauty and integrity of life, Carson continues to inspire new generations to protect the living world and all its creatures.

You may be wondering what any of this has to do with the Arctic. As I was to discover when I got down to work in my newly elected role as president of ICC Canada, toxins affect life in the Arctic all too powerfully.

When I joined the ICC, a number of initiatives were already under way. I had to bring myself up to speed quickly to contribute to the projects we were doing in Russia and Belize. I also got a crash course in the environmental work that the ICC had taken on since the organization was created.

It was nearly by accident that scientists discovered that the Arctic had a problem with a group of contaminants called persistent organic pollutants. The Arctic captures the imagination of many not just because of its majesty but also because of its perceived pristine environment. A group of scientists in the 1980s studying pollution in the Great Lakes also assumed the Arctic to be an almost-pure environment. As they embarked on their Great Lakes Action Plan studies, they were looking for an ecological baseline to compare with toxin levels in the biota of the Great Lakes. They started by testing the icy waters above the sixtieth parallel, far from the factories and smokestacks of the South, then moved on to test for toxins in the breast milk of nursing Inuit mothers and in the cord blood of Inuit infants. What they found was shocking. The Arctic could *not* provide a toxin-free baseline for their studies. What's more, the levels of contaminants in the marine and human life of the Arctic were five to ten times higher than they were in the Great Lakes biota. How was this possible when pesticides weren't used in the North and there was no industry spewing pollution and toxins?

It didn't take the scientists long to figure it out. POPs are largely synthetic chemicals used in pesticides, herbicides, industrial processes, and manufacturing. They share a number of characteristics: they are extremely volatile, which means that they evaporate in warm air and condense easily in cold air, and their molecules do not break down readily, having a low water solubility, but are highly soluble in fat.

The warmer the climate, the more likely POPs are to evaporate, allowing them to travel on wind currents. The colder the climate, the more likely they are to condense, and therefore travel through the atmosphere and water (snow and ice) systems. Because these molecules are highly resistant to breakdown, they can go through many seasonal cycles, evaporating in the summer and condensing in the winter, moving northward around the world on the planet's weather systems. But they don't get evenly distributed—they move relentlessly from warm to cold. They are a bit like a pinball. They can bounce around for a long time, but eventually there is only one place they end up: the coldest climates on earth. In other words, Arctic land and waters. This process is also called a "grasshopper effect." It's estimated that fully 80 percent of the pollutants found in the Canadian Arctic come from outside of Canada.

While POPs' molecular stability (their "persistence") and their tendency to evaporate and condense mean that they naturally migrate to the Arctic (and to higher elevations, like mountaintops), another property complicates their presence in this ecosystem. POPs bioaccumulate and biomagnify in the environment. That is, because they don't easily break down and because they are soluble in fat, they accumulate in living creatures as they move up the food chain. The higher the predator, the higher the concentration of POPs in its tissues. Marine mammals, like narwhal, beluga, walrus, and seals, are likely to have much higher levels of POPs in their blubber, for example, than fish. And animals with a high concentration of fat (which whales, seals, and walruses all have to help them survive frigid Arctic waters) will store more POPs as well.

Caribou and land animals tend to have lower levels because they have less fatty tissue. And finally, because these chemicals are persistent, they remain in the fat as an animal grows. The older the animal gets, the more POPs it has ingested, the higher the concentration of POPs in its system. In short, marine mammals, the core of our country food, food we have relied on for millennia, act like global conveyor belts funneling high concentrations of toxins into our Inuit populations.

In fact, because our diet is relatively limited in our northern communities, we Inuit are disproportionately exposed compared to people living in the South, and even compared to other Aboriginal peoples of the circumpolar regions. Inuit are coastal people and rely heavily on marine mammals. And studies conducted in the eighties and after showed that, in our country food, polychlorinated biphenyls (PCBs), chlorinated pesticides, and heavy metals like mercury, lead, and cadmium were all at levels higher than recommended.

Added to the danger is the fact that the chemicals are tasteless and invisible and have no immediate effects—making it impossible for anyone to know how much exposure they are getting. And while all the POPs that were originally studied (dubbed the "dirty dozen") are known to pose health risks, most health studies looked at the effects of only one chemical in isolation, at very high exposures, for a short period of time. Yet we were being exposed to a cocktail of many chemicals, in a range of exposures, over a long period of time. Further, our populations were small, and this made it hard to link an exposure to specific results. We did know, however, that the levels of certain POPs in our food chain were well above those recommended by most health authorities, and that high levels, higher than in women anywhere else in the world, were found in the breast milk of our nursing mothers. In fact, studies confirmed the presence of more than two hundred chemicals, including DDT, PCBs, dioxin, lead, mercury, toluene, benzene, and xylene, in the milk. Toxins were also found in the cord blood of our babies. The existing research suggested that our babies were at higher risk for serious ear infections, impaired neurological development and intellectual functioning, and lower birth weight. POPs put older community members at risk for diseases such as cancer, especially breast cancer, and osteoporosis. Though POPs are a truly global issue—the U.S. State Department has called them "one of the greatest environmental challenges the world faces"—they disproportionately affect us Inuit. And the sad irony is that while our risk is so elevated, we have received no benefit—we have never used POPs for improved agriculture or to prevent malaria—from these toxins.

Scientists and researchers of NGOs, governments, industry, and health agencies had been releasing data and reports on POPs and Inuit food and health since the mid-eighties. During that time Mary Simon, the international president of the ICC, and ICC Canada president Minnie Grey (whom I had been on the Nunavik Education Task Force with) had been working with others internationally to address the crisis, and had also been working locally to help our community understand the problem and to deal with their health concerns. But while others had been grappling with this issue for years, much of the POPs research was new to me. I had heard that there were environmental concerns in the Arctic, but during the years these initial discoveries had been reported, I was living in the South, preoccupied both with caring for my young family and with my work on education and youth. But like so many other problems that were affecting our Inuit communities, this one now hit very close to home for me. Of course, being an avid country-food eater myself, I had to wonder about my own exposure to POPs. And I was worried about how my own family's health might have been affected.

But the threat to our country food struck me at a deeply visceral, emotional level as well. Many in the community were wondering whether they should give up breastfeeding their babies or eating marine mammals entirely. The thought left me shaken. Not being able to eat *muttaq*, seal, caribou, or ptarmigan during my eight years away had been one of the hardest things about leaving my childhood home behind. My return to Kuujjuaq at eighteen had also been a return to a diet that nourished me not only physically but spiritually as well. The animals that are our country food connect us to the water and the land, to the "source" of our life, to God. Often when I prepared country food, my hands fully covered in blood, I would think that those who garden in the South must feel the same, their hands covered in the soil in which their vegetables grow. Source is source, whether it is the blood of the animals we hunt and eat, or the soil in which we grow our food. All comes from the same place.

The preparation of this food, plucking and gutting geese, cutting up caribou or seal, also links us to our forebears, to our families, to our community. (I have often experienced this connection. Whenever I plucked geese or cut up my caribou or fish in the years immediately after I lost my sister, my mother, and my aunt, I would feel their spirits beside me as I worked. Feelings of grief and loss would immediately surface. My tears would become a cleansing spiritual experience as I slowly let them go.) As an adult, living in the South again, I had brought country food back with me after every trip to the North. Friends and relatives regularly sent me some of their bounty as well. Sharing country food is a deeply ingrained

Inuit tradition. I knew that losing this food, this tradition, this spiritual connection to my world would be as devastating for me as those early days in Blanche and Churchill had been. And I knew I was not alone in feeling this way. For Inuit, not eating country food is like Italians being told they can no longer eat pasta, or Americans being deprived of their burgers and fries. This was not just a matter of losing something that we loved. Though, of course, it was that too.

Most devastating was that this food that nourished us in our cold environment, that we cherished, that had held families and communities together for generations, had turned out to be toxic. The prospect of losing our food turned our world upside down. What had saved us in our Arctic environment was now poisoning us. Now anything around us could be a threat. Now anything could be taken away from us by distant strangers. It is impossible to describe just how vulnerable this made us feel. And we didn't just *feel* vulnerable. We *were* vulnerable.

As I learned more about how POPs were contaminating our food and creating a health crisis, I became increasingly passionate about tackling the issue that posed such a threat to the people and the culture I had been mandated to protect. Luckily, as president of ICC Canada, I was able to join the global effort. Indeed, my work would propel me onto the global stage almost overnight.

THE INTERNATIONAL WORK on POPs reduction had started years before I became involved with the ICC. In 1991, representatives of the eight Arctic states that would eventually become the Arctic Council—Canada, Russia, the United States, Sweden, Norway, Finland, Denmark, and Iceland—came together to discuss how to reduce the effects of transboundary

POPs on our environment. Out of these discussions, the states adopted an Arctic Environmental Protection Strategy (AEPS), which included the Arctic Monitoring and Assessment Programme (AMAP). AMAP would keep close watch on Arctic environmental pollution issues, including POPs.

Canada initially chaired the AEPS work, and in 1994 established its own Northern Contaminants Program (NCP) under the federal Department of Indian Affairs and Northern Development to monitor and study pollutants in the Arctic. ICC Canada soon got involved with the process, attending meetings at the regional, national, and international level. To move this from a national or regional Arctic concern to an international issue, however, Canadian and other Arctic national representatives had to find an international home for POPs reduction negotiations. The only international mechanism at the time that could initiate such negotiations was the United Nations Economic Commission for Europe (UNECE), which included Europe, Russia, Canada, and the United States. UNECE had identified the threat of POPs after Canada and Sweden raised concerns as far back as the late eighties. These two nations believed that the existing Convention on Long-Range Transboundary Air Pollution (LRTAP), which dealt mainly with acidification (acid rain), could be the home for future work on a POPs protocol—a document, legally binding for the countries involved in UNECE, that would identify the pollutants of greatest concern and outline steps to reduce their use and their spread, and that could be the basis for an international agreement on POPs control. UNECE agreed to undertake these negotiations for the development of this regional protocol, which would happen over a number of years. Eventually, the LRTAP process would form the foundation for the global negotiations on POPs regulation that the United Nations Environment Programme (UNEP) would take on.

UNEP's first order of business was to hold a series of technical meetings to describe the problem and come to scientific agreement on the characteristics of the chemicals of concern. In May 1995, the UNEP governing council called for an international assessment of twelve POPs, the dirty dozen. Through a series of meetings, it was agreed that a global protocol on POPs was necessary, and a series of five Intergovernmental Negotiating Committee (INC) meetings were proposed, the first to be hosted by Canada in Montreal in June 1998.

However, the official negotiations for the chemicals that would be included in the protocol began two years before this first INC meeting, in 1996. The first meeting at which the ICC was officially represented under the Canadian flag took place in Geneva in the summer of 1997. Stephanie Meakin, an environmental biologist under contract with our ICC Canada office, attended as our representative.

UNEP is very public in its recognition that sustainable development is dependent on business and industry participation, co-operation, and initiative, so representatives from the private sector are always involved in the UNEP meetings and negotiations. Given that businesses and industries will have to implement changes to their operations if the member countries ratify an agreement to reduce toxic emissions or restrict the use of certain chemicals and, furthermore, may be able to suggest alternative chemicals or processes, this makes sense. But these meetings were *dominated* by industry groups. Their representatives saw the negotiations as a real threat to the then-unimpeded trade and movement of chemicals globally—a billion-dollar industry. The only environmental group present was Greenpeace. The Inuit Circumpolar Council participated as a non-governmental organization, lobbying for recognition that the Arctic and its population were particularly vulnerable to POPs.

I entered this public international effort as a politician determined to protect our people from these dangerous chemicals, but also as a concerned mother—and a grandmother. Near the end of 1996, I learned that my daughter, Sylvia, was pregnant. She gave birth to a son, my first grandchild, in July 1997. Here was another generation that I now knew was at risk. As Sylvia nursed her newborn, I was struck by the injustice of what we were facing. No mother anywhere in the world should have to worry about whether her breast milk will poison her child. Ultimately, the POPs issue was not about politics, but rather about families, parents, children, and grandchildren—and our right to lead our lives and continue the strong traditions of our hunting culture. Traditional Inuit wisdom, the powerful teachings of the land and the hunt, and our country food were all vital to confronting the historical problems and the rapid changes that arrived with the modern world. I wanted to see my grandson, and the grandsons of many, grow up and live in the Inuit hunting and food-sharing culture that had sustained me during my upbringing, and that had fulfilled the physical and spiritual needs of Inuit for generations before.

Yet our fight was a complicated one. Many mothers in tropical climates relied on the use of the pesticide DDT (one of the dirty dozen) to kill mosquitoes and therefore protect *their* babies and children from malaria. I wanted to create a healthier world for mothers *everywhere*. And while the health of families in the South needed to be protected too, it was becoming clear that what was happening in the Arctic would eventually happen to everyone else. As I often say, there may be only 160,000 Inuit in the entire world, but the Arctic is the barometer of the health of the planet. If the Arctic is poisoned, so are we all.

The scientists and environmental experts were bringing the data and the research to the discussions. In particular, the Canadian Northern Contaminants Program and the Arctic Council's AMAP were generating information and reports that paved the ground for the ICC to build upon in our later policy work. In 1997, in fact, they were to release the most comprehensive regional contaminants assessment report in the world, *Arctic Pollution Issues*, which recommended immediate international action on the issues of POPs and transboundary pollutants.

But we needed to bring the human dimension to the scientific, chemical-related discussion—to put a human face on the POPs issue. The human population of the North had a tight bond with the land and the animals—the lives of thousands of people were at stake. This was not a North–South issue, but rather a global issue. This crisis was not only harming our people but also adding to existing economic and social problems. Everything was connected.

I came into the process in 1996, just as the Arctic Council, made up of representatives from the eight countries that had originally met to discuss toxins in the Arctic, was officially established, and INC meetings about POPs in Montreal were set to start. I made a series of presentations at various meetings and delivered speeches at conferences on behalf of ICC Canada.

In the fall of 1996, I spoke at the Canadian Polar Commission's conference on contaminants in the environment and human health, which was held in Iqaluit. Later that year, I gave that same speech at the International Union for

Conservation of Nature (IUCN) in Montreal. During those talks, I told the audience that the issue of contaminants could not be treated separately as an issue unto its own. It could not be isolated from the many other challenges that we were faced with as a people. It was yet one more difficulty placed on us from the outside, adding to all the other historical traumas that had come into play for us in the fifties, sixties, and seventies. Now, on top of everything else we were grappling with in our communities, we had to deal with these toxins in our food and in our bodies. I tried to capture what I had learned from documents of past ICC work and from others at ICC meetings and assemblies who had been talking about how communities were reacting to this newest blow.

Minnie Grey has described to me just how harrowing the news of POPs was when it first reached our communities in the mid- to late eighties. She told me that panic spread when the results of the work of Dr. Harriet Kuhnlein from McGill University and Dr. Eric Dewailly and his team from Laval University appeared in local and international media. The research teams had found PCBs in the breast milk of nursing mothers from Nunavik. PCBs have long been linked to cancer, especially breast cancer. Studies have also shown that infants exposed to PCBs have higher than average rates of impaired motor and cognitive abilities and compromised immune systems. Nursing mothers were devastated to think that they might be passing on poison to their babies. Many felt that they should quit eating country foods. Indeed, many stopped breastfeeding and replaced breast milk with evaporated milk—all that was available in the small Inuit communities.

Yet it was not feasible for these mothers and their families to give up the food that we harvested from the land. The selection of southern food available in our stores was limited. Fresh fruits, vegetables, and meat were hard to find. And even when this food was available, it was extraordinarily expensive. (For example, a 2014 Statistics Canada survey of food prices in the Arctic reported that a box of soda crackers in Clyde River would set you back $11.49; a kilogram of stewing beef in Pannituuq cost over $20.00; and a kilogram of carrots in Pond Inlet cost $8.09. Not surprisingly, in 2014 the Council of Canadian Academies reported that Inuit go hungry more than any other Indigenous group in Canada, or, in fact, in all of North America.) We could not simply drop seal meat, *muttaq*, and walrus from our diets and replace it with a well-balanced Mediterranean spread. The Inuit community felt ambushed by this dilemma.

Minnie and fellow Inuit leaders knew they had to find ways to share information and calm fears. But it was a daunting task. Minnie, who is fluent in English and Inuktitut, and interpreter and translator Martha Kauki realized that they didn't even have the language with which to discuss the issue with our population. There is no word in Inuktitut for *toxins* or *pollution*, never mind *POPs*. They settled on *sukkunartuit*, the Inuktitut word meaning "harmful substances."

Meanwhile, Nunavik health authorities and some leaders, with the assistance of Eric Dewailly and his team from the Public Health Research Unit of the Laval University Medical Research Centre, decided to set up a committee that would communicate the scientific information about PCBs and their sources in layman's terms. According to Minnie, even the apparently simple act of naming the group highlighted how challenging it was going to be to share the important information now available without generating more panic. The first name chosen, the "PCB Committee," created confusion and alarm in the communities. Then in 1992, Dr. Dewailly, along with his team, requested a meeting with Minnie, who was now the executive director of the Ungava Tulattavik Health Centre. Following a long discussion of the findings, it was decided that the PCB Committee would be renamed the Foods and Contaminants Committee. They then set up a plan to communicate information and provide guidance more effectively. But the committee soon realized that the word "contaminants" in their title was still distressing rather than reassuring to people. They renamed the committee for a third time, as the more positive Nunavik Nutrition and Health Committee, and focused on providing information about contaminants in the food chain, while stressing the importance of the traditional Inuit diet. They advised nursing mothers that our country diet, despite the presence of toxins, was still far better for them than store-bought, processed foods, and that breastfeeding was still the best thing for their babies. The committee made posters and videos and appeared on radio shows to get these messages out. (The committee is still operating, leading many research projects and giving advisories about food safety for babies, pregnant women, and women of child-bearing years. The Arctic char program, which gives out free fish, a less toxic food choice than marine mammals, to expectant and nursing mothers, is one program that came out of this committee.)

Of course, the panic and upset caused by the discovery of POPs in the Arctic wasn't limited to the Inuit communities of Nunavik. Research was happening in Nunavut, as well. According to Harriet Kuhnlein, who led McGill University's PCB research in Qikiqtarjuaq, Nunavut (known as Broughton Island, Northwest Territories, at the time), the initial research results her team had gathered were not meant to be made public before the mayor of Qikiqtarjuaq was fully informed. Somehow, however, the news broke and, in Harriet's words, created a nightmare. A public meeting was held with the mayor and his council, community members, and media, including CTV. Many people were extremely upset, perceiving the report about PCBs in their food as yet another government directive telling them how to live their lives—in this case, what to eat and what not to eat. Some felt that it was a plot to get them off the land by suggesting that their food was poison.

After incidents like this, the ICC, the national Inuit Tapiriit Kanatami (ITK), the Dene Nation, the Council of Yukon First Nations, and others insisted that they be given research results and be allowed to report back to their communities before anything was shared with the media or the larger public. In 1991, in response to these concerns about human exposure to elevated levels of contaminants in the traditional diets of northern Aboriginal peoples, the Northern Contaminants Program, led by Garth Bangay and later David Stone, within the federal Department of Indian Affairs and Northern Development, was established. The NCP undertook research into effective communication of the issues, as it had experience in the area from its research contributions to AMAP, which has been headed by Lars Otto Reierson from Norway since its creation. In 1992 the Centre for Indigenous Peoples' Nutrition and Environment (CINE) was also created in response to a need expressed by Aboriginal peoples for research and education to address their concerns about their traditional food systems. In 1993, aptly headed by Harriet Kuhnlein, CINE found its physical space within McGill. The CINE governing board included representatives from the Assembly of First Nations, the Council of Yukon First Nations, the Dene Nation, ICC, ITK, the Northwest Territory

Métis Nation, and the Mohawk Council of Kahnawake. Part of CINE's mandate was to include Aboriginal peoples in the research and education of their communities related to traditional food and nutrition. CINE worked with ITK, community organizations, and the media to share nutritional information for pregnant and nursing mothers and to give advice about the quantities of various country foods that could safely be consumed. Harriet Kuhnlein and her organization continued to stress that the benefits of a traditional diet far outweighed the risks. CINE played a key role during that period of uncertainty and beyond. It helped our northern world to avoid feeling paralyzed by the chain of events that seemed far outside of our control.

While I tried to convey to the international groups I spoke to what I had learned about the effects of the POPs scare in our communities, I also attempted to increase their understanding of the social context. During the initial meetings of the newly formed Arctic Council, I walked the senior Arctic officials, high-level government bureaucrats representing the eight circumpolar countries, through some fundamental yet vital information about the social, health, and economic conditions our communities were facing. Many of the government representatives of these circumpolar nations, each with Indigenous peoples living in the northern part of their countries, had never ventured north to experience Arctic life as we lived it. What's more, although the ICC had been part of the UN processes for many years as a member of the United Nations Economic and Social Council (ECOSOC), we could not assume that the Arctic Council, composed largely of government bureaucrats who weren't Indigenous circumpolar people, and those in the international arena fully understood the Inuit position and perspective.

It was important to me, as it was to many fellow leaders, not to waste the potential of what this new Arctic Council could accomplish. Many people in our communities were understandably worried that the Arctic Council would become yet another institution intent on "fixing" things as it saw them, with no real connection to our lives. Showing the council what daily life was like in Arctic communities would be the best way to convince them that we needed a holistic approach to dealing with Arctic problems—no matter which institution was doing the work. My time with the Nunavik Education Task Force, as well as my work on the regional health report, had given me plenty of grassroots exposure to our Arctic realities and had provided powerful lessons in how everything was connected. I wanted the Arctic Council, and the other groups I spoke with, to be able to connect the dots too. So while I addressed the problems of POPs, and stressed that the issue of pollution and contaminants in our food was a crisis not of our own doing, I also talked about the self-induced contaminants that were afflicting our communities, in particular, tobacco, alcohol, and drugs. We were already in a vulnerable position, I said, and it had led many to substance abuse and addiction. We were struggling to make sense of it all. We needed those looking at the POPs issue to understand the entirety of our challenges, to understand what was happening on the ground in our communities. This was not just an issue of the environment or of science, but first and foremost, one of health. If the Arctic Council could have as one of its main vision statements the goal to turn around the statistics of addiction, poverty, and suicide by genuinely empowering the Indigenous communities, we would be halfway there.

IN MY FIRST FEW YEARS with the ICC, I spoke about the POPs issue in the Arctic at a dozen events, at the very least. These included meetings and conferences for the United Nations, the Conference of the Parliamentarians of the Arctic Region, the Northern Forum, the Canadian Polar Commission Conference, and Canadian Senate hearings. I also did numerous print, TV, and radio interviews, which were aired in our communities, as well as nationally and internationally. As I talked more and more, I became aware of the tricky balancing act that discussion of POPs necessitated. Fear was a topic that re-emerged frequently, for example, in the initial Arctic Council meetings on Arctic contaminants. Our northern communities were in a state of confusion and despair. Not only were we dealing with inadequate housing and health care, deficient justice and education systems, and now toxins in our food, but also, in 1993, the federal government had introduced the Canadian firearms registry. In an effort to lower gun-related crime, this new law required all firearms, including hunting rifles, to be registered with the government. This meant paying a fee and filling out paperwork, which many in our community—and many others across the country—found onerous. But our hunters also feared that they would be jailed if they didn't adhere to the gun-control law. I suspect many of us heard echoes of forced relocation, dog slaughter, and other official actions in all this. Certainly, the long-gun registry left many feeling that, once again, we had no agency or control over our lives.

I realized that my message to my community needed to pick up on the positive, reassuring work that the Nunavik Nutrition and Health Committee, NCP, and CINE were doing. They were stressing that our traditional or country foods were still the most nutritional food we Inuit could consume since, unlike Southerners, who might decide to avoid certain large fish species for fear of contaminants, we couldn't just switch to other safer fresh food. We had significantly fewer healthy choices. Cutting out our country foods would mean eating a diet of largely processed southern foods. Such a shift would create a host of physical, mental health, and financial issues. So, while there were health risks to eating our country food, eliminating it completely from our diets posed health and other risks as well, ones that were potentially even more harmful than the toxins—the benefits of the food, therefore, outweighed the risks.

At the same time as I was trying to calm and reassure our communities, and build a sense of trust and safety in the ICC leadership on this issue, my message to the international community had to ring alarm bells. The process of negotiating to change global environmental and economic policies is notoriously slow. UN treaties are years and years in the making, therefore. I had to advocate for an urgent stop to the use of the toxins. I also had to implore the bureaucratic and scientific communities to help get the message out rather than get caught up in debates about methodology and differing interpretations of research results. It was a strange balancing act—having to be reassuring at one point and alarmist at another.

The more presentations (known as "interventions" at the UN negotiating meetings) I made and meetings I attended, the more frustrated I became with the scientific squabbling that often took place. Different research institutions had different research standards, and disagreements about how to interpret the data were frequent. In particular, industry-backed researchers often seemed intent on focusing on the deficit of long-term data. The small sample sizes from the small Arctic population also led to much debate. And, of course, there was the problem that so much of the existing research on persistent organic pollutants focused on the risks that individual chemicals posed, rather than on how chemicals acted in combination. Those of us representing the Arctic communities were not snubbing the scientific data or discounting the need for further studies—especially studies that looked at the effect of the toxic cocktail floating through Arctic waters. But this work was under way, and we knew it would take time. Time we didn't have. There was ample scientific evidence that we Inuit were at risk. And we needed action *immediately* to prevent further damage to our environment.

Yet, the industry representatives at these meetings seemed more than willing to let scientific debate stall the process and imperil Inuit communities. At an Arctic Council meeting in Oslo in 1996, I became so frustrated that I told the attendees I felt history was repeating itself. Missionaries, fur traders, and governments had fought over the Arctic for decades to further their own self-interest: converting us to their religion, pressuring us to build their trade, or using us to establish their sovereignty. In the process, our well-being and our way of life were sacrificed. Here again, scientists, consultants, and lawyers were busy pushing forward their own agendas while we suffered.

Although it was aggravating at times to be repetitive in my presentations at the Arctic Council meetings, I felt the need to state again and again that the ultimate goal was to work in genuine partnership with one another to phase these contaminants out of our food chain. I knew that if our differences weren't worked out at *this* table, it would be difficult for the Arctic Council to work as a bloc and to be strong in its stance globally when the UNEP negotiations began. I pushed the idea that these conflicts and differences had to be resolved.

I wasn't suggesting that there had to be unanimous scientific agreement; however, I was clear that we Inuit were not willing to be caught between these differences. What's more, we needed the experts to provide clarity for our communities—to convey what they knew in terms that we could share with those affected in the Arctic, without confusing the data by using different standards, measurements, and so on.

At a presentation to the Secretariat of Arctic Council officials in Oslo in November 1996, before the initial INC negotiations on the POPs treaty had begun in Montreal, I spoke strongly about these concerns, and once again, I tried to put a human face on POPs debate. Right from the onset of my global work, I felt compelled to have those in power react to these issues with their hearts and not just their intellects. I still work this way today, and for me it feels like a much more effective way to make changes to a political world, which so often avoids working with the heart.

WHILE ALL OF THIS WAS GOING ON, in March 1998, my first term as ICC Canada president came to a close. Yet I felt I was just beginning. I decided that I would run again, and as no one ran against me, I was acclaimed. This elected position with ICC Canada meant that I was vicepresident for ICC International as well. The bylaws of ICC Canada had changed since 1995, and my position had moved from being a part-time, almost-volunteer role to a full-time position. I had already decided that my efforts were best spent representing our people internationally, which meant leaving the regional mandate of the Makivik Corporation. While I had enjoyed the close work with my community, this new focus felt right for me.

Since I would no longer be working at Makivik, and the

ICC headquarters were in Ottawa, I decided to move there from the West Island of Montreal. My daughter was living in Iqaluit. My son had been living with me in Montreal, but he was working as a commercial pilot and could get his own apartment. Other than my maternal concerns about leaving Eric, there was no reason not to be where the ICC was based.

Shortly after I started my second term with ICC Canada, I spoke in Montreal at the first of five international meetings that would negotiate the terms of the UN Stockholm Convention on Persistent Organic Pollutants. Over the next two and a half years, we worked to prepare a legally binding instrument for implementing international action by the year 2000 on an initial list of twelve POPs—the dirty dozen. I would attend every single one of the negotiating sessions, starting in Montreal in 1998, then continuing in Nairobi, Kenya; Geneva, Switzerland; Bonn, Germany; and Johannesburg, South Africa. I was joined at all these meetings by ICC Canada contractual employee Stephanie Meakin and strategic counsel Terry Fenge. John Buccini from Canada was elected as chair of the INC process and became a valuable friend in the work I was to undertake.

The INC meetings were huge gatherings, attended by about seven hundred people from over one hundred nations. Participants included governmental representatives from many of these countries, including Canada, the United States, Russia, Germany, Sweden, Denmark, African nations, Mexico, and many more; academics; environmental researchers; scientists; lawyers and lobbyists from manufacturing industries and their associations; environmental NGOs like Greenpeace and the World Wildlife Fund; NGOs representing Indigenous and Arctic interests, including the Canadian Arctic Indigenous People against POPs; and a variety of journalists. Each meeting stretched over several days. Attendees would gather in one huge room, with the chair on a raised dais at the front. Delegates from each country or group would make interventions, asking for changes to the draft convention document, noting what language they could accept, what phrasing they would not approve, and what they wanted added. The secretariat would note all of these proposed revisions, putting the new versions in square brackets in the draft. Small "contact" groups would then negotiate these changes outside of the plenary sessions, often actually working in the hallways late into the night. Not until the changes got unanimous approval within the group would they become part of the final convention document. When a group agreed to a change, it would present that approval in the next plenary session, and the secretariat would remove the square brackets from the text. It was an exhausting and tedious business.

The ultimate aim of the treaty was to eliminate persistent organic pollutants at their source. But to go about writing the document was a complicated matter, with many competing interests to consider. Scientists debated which chemicals should be identified in the protocol. Industrial groups had concerns about the costs and the disruption that elimination of these chemicals might place on manufacturers. While many of the listed chemicals were already being phased out, chemicals like DDT represented a billion-dollar industry. What's more, those in the chemical business feared that introducing global restrictions on the currently identified POPs could set a precedent down the line that would affect other chemicals that were still in use, and they fought hard against any regulation. Government representatives often focused on the effects that POPs reduction would have on their economies, as well as their environments, while the Arctic communities and environmental NGOs tried to move the spotlight onto long-range human and environmental costs. A great deal of work also went into identifying realistic reduction percentages and target dates. (For a definitive history of these negotiations and the many parties involved, see *Northern Lights against POPs: Combatting Toxic Threats in the Arctic* edited by David Leonard Downie and Terry Fenge.)

(At these meetings, ICC representatives were part of the Canadian Arctic Indigenous Peoples Against POPs [CAIPAP]. CAIPAP had been formed with funding from Environment Canada and the Department of Indian Affairs and Northern Development. Its mandate was to press for a global POPs convention that would be comprehensive, rigorously implemented, verifiable, and a source for POPs–related public health information. CAIPAP was supported technically by Stephanie Meakin, under contract to ICC, and included Aboriginal representatives Eric Loring from Inuit Tapiriit Kanatami, an expert on contaminants in Arctic communities; Cindy Dickson from the Council of Yukon First Nations; Carole Mills, a Dene woman who worked with the Northern Contaminants Program in the Northwest Territories; and me, acting as spokesperson for the group. The CAIPAP team worked closely with David Stone and the Canadian Northern Contaminants Program, and relied on the legal counsel of Anne Daniels and the scientific support of Russel Shearer and Siu-Ling Han from the Northern Contaminants Program, Andrew Gilman from Health Canada, Derek Muir from Environment Canada, and Robie MacDonald and Gary Stern from Fisheries and Oceans Canada, to name just a few. We eventually expanded our associations to include the Indigenous Environmental Network headed by Tom Goldtooth of the United States Physicians for

Social Responsibility, World Wide Fund for Nature, and, in particular, the International POPs Elimination Network [IPEN], which brings together numerous organizations from around the globe.)

From the onset of these negotiations, one of the big challenges those of us with ICC Canada and CAIPAP faced was trying to get our Canadian government's head of delegation, Steve Hart from Environment Canada, to accept us as a strong component to our government's strategy (which, like that of every other country involved, was to eliminate the toxins at their source, not just manage them). In both the Arctic Council work and now the UNEP negotiations, those of us on the ICC had been encouraged by the participation and support of a number of federal Members of Parliament (MPs). Clifford Lincoln, MP for Lachine–Lac-Saint Louis, had spoken with great understanding about the plight of Northern Indigenous peoples at the inauguration of the Arctic Council. He, along with MPs Karen Kraft Sloan and Charles Caccia, now acted as overseers of the Canadian delegation for the UNEP POPs negotiations. MPs Rick Laliberte, John Herron, and other members of the House of Commons Standing Committee on Environment and Sustainable Development had also been very supportive. And Christine Stewart, the then Canadian minister of the environment, delivered a strong opening speech to the Montreal meeting, in which she talked about the high levels of POPs in the breast milk of Inuit mothers.

But despite the support of these politicians, government agencies are used to working on their own, and it can take some effort to find a helpful and complementary relationship. Part of the ICC's challenge was that we were an NGO, a non-profit, voluntary citizens' group. As such, we had been invited by the Canadian government to be part of their delegation, but we weren't necessarily negotiating in the trenches with them. We had observer's status at the United Nations, which meant that we could speak independently from the floor during the negotiations (with the recognition of the INC chair), but we always had to be mindful that our funding sources included the government of Canada. And we knew that the most effective way to get our message across was if it was represented in the Canadian government's speaking points as well.

At the outset of the negotiations, we designated Carole Mills, from CAIPAP, to attend the Canadian delegation meetings each morning and put forth our issues and needs to the federal negotiators. Carole, who was knowledgeable on the subject of contaminants, was well placed to act as our liaison with the Canadian government delegation, but at first, Steve Hart seemed distrustful of this process and uncertain about what we could contribute. At one point he said, "I guess it's better to have a skunk in the tent spraying outward than outside spraying in." We weren't pleased to be characterized as skunks. It was clearly going to be a bit of a struggle to find a way to work with our government. And this hurdle wasn't always helped by the personalities involved.

In between the sessions at the INC meetings, ICC Canada's strategic counsel, Terry Fenge, and biologist Stephanie Meakin worked with the Canadian delegation to ensure their goals were in line with our ultimate aim: to make country food safe for all Inuit and other Aboriginal peoples. A native of Britain, Terry was a politically astute policy expert with two degrees in geography and a Ph.D. in regional planning and resource development from the Faculty of Environmental Studies at the University of Waterloo, and a dry, biting sense of humor. I had hired him personally and for eight years relied heavily on his keen ability to see the lay of the land and address the many obstacles that were being thrown our way. He was particularly good at taking on the industry interests, whether it was chemical manufacturers or industrial lobbyists, who were heavily represented at the INC meetings. (The Canadian Chemical Producers' Association was at every INC session.)

But Terry could also be a bull in a china shop. He had a strong sense of confidence and very high standards. And he didn't bother to mince words when something failed to live up to them. As a result, he often ended up rubbing people the wrong way, even people with whom he worked closely.

One of these people, unfortunately, was the head of the Canadian delegation. Terry felt that Steve Hart did not appreciate the importance of the public health impact of the POPs issue and wasn't representing this concern in Canada's policy position as strongly as he should have been. Both Terry and I suspected that the fact that Steve Hart had worked for many years with industry and, as far as we knew, had never worked with Aboriginal peoples prevented him from fully appreciating the Inuit position. And he certainly misjudged my access to his political and civil service bosses. In the beginning of the INC work, his offhanded approach to the urgent health issues frustrated us. Terry came to me and said he wanted to write a private letter to Steve Hart asking him to step down and allow someone with greater sensitivity to lead Canada's delegation. On his own letterhead and with my approval, he did so. Shortly after that, I received a call from Hart, requesting a private meeting with me in my Ottawa office. While I agreed to the meeting, I explained that, as my strategic counsel, Terry Fenge would be there as well. Steve Hart was not pleased, although he did come. He arrived with someone from the foreign affairs office, who was part of the negotiating delegation from Canada. We had gotten to know both of these men quite well.

We began our meeting on a positive note. Although Terry had asked for Steve's resignation, I did not raise this, but instead focused on how we could better our partnership. They had started to consider the ICC a serious component to the global negotiations, having now seen us in action at a number of meetings. And they were beginning to recognize that we Inuit could be bridge-builders between the affected North and the producer South. As Inuit and as Canadians, we had a stake in both worlds. We also brought a sense of moral authority to the negotiations—our threatened health and lives were really reflections of the future of the planet as a whole. Yet, when we came to discuss the tensions in our partnership, Steve Hart was clear about what, or rather who, was the problem.

"We do not have a problem with any of this, Sheila. The problem lies with him." He pointed his finger directly at Terry. "This man is crass. He's arrogant. He's abrasive, and he's obnoxious."

Terry was sitting beside me. I looked at him, putting up my hand. "Terry, don't say a word," I said.

I proceeded to tell Steve Hart that I was aware of everything that Terry Fenge did on my behalf, and he, on all occasions, had my blessing to say and do what was needed to raise awareness of our issues. Terry worked for me. He was doing what he had to do to make sure we were heard. Steve Hart got that message. Terry would later say, in reflecting on our struggle to be included in the Canadian delegates' strategy, "I don't think of Steve Hart as a source of our problems, but rather the expression of them."

Just as Steve left my office, both Terry and I stood up. Terry uncharacteristically gave me a big hug, and I said, "How dare he carry on like that! I'm the only one who can call you crass, arrogant, abrasive, and obnoxious!" We both had a good laugh about that.

Later on, it became clear that Steve Hart had understood the value I placed on Terry. The incident was never discussed again, and we went on to have a fruitful partnership with the head of the delegation and with many in Environment Canada, the Department of Indian Affairs and Northern Development, the Department of Foreign Affairs, and others. (All members of the Canadian delegation played an important role during the negotiations, but it was David Stone of the Northern Contaminants Program, within Indian Affairs and Northern Development, and his international role with the Arctic Monitoring and Assessment Programme, as well as Ken McCartney of Foreign Affairs, who were most instrumental to me during this work. David came to be our closest ally within the federal system, and I am forever grateful for his ability to see beyond his bureaucratic role.)

AT MY PRESENTATIONS at the INC meetings, I was always conscious that I represented the voice of Inuit of Greenland, Alaska, Russia, and Canada. As such, I needed to do three things when I spoke publicly about POPs. First, I had to show that I was well informed about the science. To achieve this, biologist Stephanie Meakin kept me abreast of the current research. (She was also either pregnant or nursing a newborn at all these sessions, lending powerful symbolism to our message.) Second, I needed to ensure, at the INC meetings and beyond, that I spoke to audiences with influence and access to decision-makers, people who could move our message forward. And third, I needed to express the moral authority of Indigenous peoples everywhere in the circumpolar world on this issue.

But interestingly, it was not a speech but a gesture I made that may have been my most effective contribution to our cause. In January 1999 at the Nairobi conference, during an evening reception sponsored by the International POPs Elimination Network, I presented an Inuit soapstone carving of a mother and child by Lucy Meeko, a fellow Inuk woman from my region of Nunavik, to Klaus Toepfer, the director general of the United Nations Environment Programme. He, in turn, presented the carving to the chair of the INC sessions, John Buccini, who, with his calm, organized, and inclusive approach, always reminded me of our Inuit elders. John made sure that the sculpture sat in front of him for the rest of the negotiations. He told me later that seeing the mother-and-child carving had been a powerful reminder of why we were all there, and that he looked to it whenever he needed energy and strength.

In my remarks to the Nairobi plenary intervention, I explained the connection that was so well captured by the carving:

> As we put our babies to our breasts, we feed them a noxious chemical cocktail that foreshadows neurological disorders, cancer, kidney failure, reproductive dysfunction, et cetera. Many Inuit mothers, far from areas where POPs are made and used, hesitate to breastfeed their babies. This is a wake-up call to the world.
>
> Here in Nairobi I have met women from many African countries who worry greatly about the effects of pesticides on their children. We have a common cause. A coalition of citizens, particularly women, from all

regions of the globe is forming to demand that you, the governments of the world, take concrete and effective action to rid us of the POPs' threat to our children.

WHILE MY REMARKS in Montreal and Nairobi seemed well received, by the end of the first day of the negotiating sessions in Geneva, it was clear that our message faced some serious objections. Several NGOs, in particular one called the Malaria Project, had made a number of presentations that stressed the importance of DDT in preventing malaria outbreaks. Dr. Amir Attaran of the Malaria Project suggested that the elimination of POPs, including DDT, would signal the deaths of thousands, and perhaps millions, of people. He even used images of jumbo jets to demonstrate the numbers of people who would die. All I could think of was how our tiny Inuit community of 155,000 would compete with the numbers in those jets. The DDT issue was creating quite a stir among the countries in attendance, and the ICC team feared that given enough traction, this issue had the potential to derail the entire process.

Terry and Stephanie helped draft my speeches for each presentation, but I always finalized them only after we had arrived at the meeting and had a chance to gauge the mood of the sessions. I was adamant about adding the "heart" of the matter to the political and scientific words, and in grounding each speech in my culture and spirit. This time, we knew we had to focus not on the numbers, but on the humanity of our story in a much larger sense. If the debate became a numbers game, we as Inuit and Aboriginal peoples of the circumpolar world would surely lose. We did empathize with people who felt that they needed DDT to preserve life and health, but we wanted the delegation to see the DDT issue as one involving the long-term survival of entire peoples, rather than as an either/or dichotomy.

In my speech, I told the negotiators that in the circumpolar world, we know about disease and death: whole families and communities had been virtually annihilated by smallpox in the last century, and some Indigenous peoples in Russia now stood on the brink of extinction. POPs elimination was an issue that concerned the survival of Arctic peoples, and indeed the health of all people around the world. "A poisoned Inuk child, a poisoned Arctic, and a poisoned planet are all one and the same," I said.

I tried to convey how difficult it was for me as a mother and grandmother to think that a mother in the Arctic had to worry about contaminants in the life-giving milk she fed her infant. At the same time, I could not accept that a mother in the South had no choice but to keep using these very chemicals to protect her babies from diseases such as malaria. At this convention and at every meeting we attended, the ICC was pressing for the identification and use of *alternatives* to POPs to safeguard the health of everyone—North and South. We were all in this together, and we had to ensure that cost-effective alternatives to POPs were made available to the developing world. I also stressed that we knew by now that if the world were to perpetually manage these chemicals and not aim to eliminate them, our country food would not become any healthier to eat. Not in my lifetime anyway. We had to remember that POPs were persistent and could accumulate for many decades. Even if the taps were shut off today, it would take fifty years to clean up the Arctic sink. I believe our message resonated and softened the North versus South divide in the contaminants issue.

Next came Bonn, Germany, for the fourth negotiating session. At the opening reception, hosted by the German government, Inuit country food was served, and a Canadian Inuit drum-dancing and throat-singing group, Aqsarniit (the Northern Lights), and the Tagish Nation Dancers (which included Tagish and Tlingit artists) performed. My daughter, Sylvia, and Madeleine Allakariallak showcased Inuit throat singing for the crowd. It was a proud moment for me.

(As a teenager, Sylvia had loved to dance, just as her mother did before her. When the accident that shattered her leg happened, she'd been involved in modern dance in Montreal. It took her several years to get back on stage, but when she finally did, it was to model Inuit designs at a Toronto event showcasing Inuit culture, called "The Spirit of the Arctic." There she met a choreographer, Alejandro Ronceria, originally from Colombia, who helped to reignite her love of dance. They worked together at many venues after that, nationally and internationally. Her newfound calling incorporated Inuit performing arts, including drum dancing and throat singing, which have continued to be the focus of her work. Her dancing and singing have taken her to many parts of the world, and there would be many times in the years to come when we'd both be invited to the same global events, sometimes with Sylvia getting the invitation before me.)

If having Aboriginal artists entertain the participants was a positive note, Bonn also presented us with some unsettling negative feedback. During our time there, we began to hear delegates say that the Arctic was getting too much attention—in the media, at the conventions, everywhere—at the expense of other people or groups. Alarm bells started to go off for us. We scurried to confront these responses and to refocus the meeting on the heart of the matter—people from the Arctic to

Mexico and India, as well as everyone in between, were at risk and would either bear the burden of a weak agreement or reap the benefits of a good global convention.

This sort of misunderstanding followed us as the process continued. Between the Bonn meeting and the session in Johannesburg, South Africa, the UNEP communication team and BBC World television crews had come to Iqaluit to film the ways that our life was being affected by POPs. The world is generally much more aware of the Arctic's wildlife than its people. We hoped that the film might address this gap. The video was subsequently shown several times before Johannesburg, and was aired in the corridors of the UNEP Johannesburg sessions. One UNEP employee criticized a scene that showed a group of us eating raw country food. It was surprising that through all the INC meetings, we had to continue to educate the global negotiators that Inuit still lived on the land—that this was not a historical way of life but a modern one. We noted that Inuit were truly among some of the last hunting-and-gathering Indigenous peoples in the world, and this was worth saving. I incorporated this statement into an intervention I gave the next day, reminding the delegation that part of our deep cultural affinity was that we Inuit eat what we hunt. I added that I hoped the delegation would continue to ensure that our health and culture were protected.

During the last negotiating session in Johannesburg, tensions were high. The recently concluded United Nations Framework Convention on Climate Change in The Hague was discouraging. While the POPs negotiations were concluding, the climate change negotiations were failing. Economic expediency and political clannishness had prevented the group in The Hague from settling on ambitious, binding targets. Part of the problem with the negotiations may have been that the

NGOs and the Indigenous peoples' organizations had little access and therefore little input into the process. At the sessions in Johannesburg, we heard through some of our channels and the media that the Hague talks were breaking down. We didn't want that to happen here. During our last POPs session, buttons with "~~The Hague~~" printed on them were circulated and prominently displayed on lapels. With a renewed sense of urgency, our ICC team continued to press for the guarantee of significant funding and technical help for the developing countries. We knew that without financial assistance to help those countries find alternatives to POPs such as DDT, Inuit health would continue to be in danger.

During all the UNEP sessions on POPs, the ICC Canada team worked closely with a number of organizations, including IPEN. This influential and well-organized group of NGOs, which included a strong presence from Greenpeace and the World Wildlife Fund, brought a collective environmental voice to the table. Terry and Stephanie maintained close connections with the group to ensure that we could tap into some of their rich resources, such as the science they had collected on POPs and their media savvy. But despite the many connections that strengthened our voice, we were also conscious that as advocates for Indigenous peoples, the ICC team needed to maintain its independence. We were often invited to join forces with other groups for media events. Perhaps it was my introverted nature or my upbringing and culture, but I just didn't feel that donning costumes and waving signs at theatrical protests was right for us. We valued the media smarts of organizations like Greenpeace, but this was not our way. What's more, we were not typical environmentalists. Our focus was on maintaining our health and way of life—including the hunting of whales and seals, which some of these groups were actively working against. If fact, the anti-sealing and anti-whaling campaigns had created much bad blood between the environmentalists and our community, which I had to respect. Now in this work on POPs, I had to be cautious not to sleep with the enemy, so to speak.

Even within our small ICC team, I was mindful of maintaining my Inuit voice. Terry often wrote early drafts of my speeches, and he loved to start them off with "I am *delighted* to be here." Whenever I read that, it struck me as a markedly British way of expressing oneself. "Terry," I'd say, "I would *never* say that." I was deeply involved in all levels of our public expression, keeping my vision and my style as an Inuk woman. I never delivered a speech unless it used my own words and energy. This wasn't an egoistic approach, but rather my way of ensuring that my representation of our people was as authentic as it could be. And my efforts in this were helped by working closely with other Aboriginal leaders, like Shirley Adamson, John Burdeck, and Chief Robert Charlie of the Yukon First Nations; Will Mayo of the Tanana Tribe of the Alaska interior; and our CAIPAP members.

That said, my close work with Terry and Stephanie was a rewriting of history in some small way—a good example of an Inuk leader and *qallunaat* advisors working together to improve Inuit circumstances while preserving and honoring our way of life.

During each of the POPs negotiations, I also made some extraordinary connections with others from around the world. When I heard that we would be going to Johannesburg for the fourth negotiating session of the treaty, I wondered if I might be able to meet Nelson Mandela. I started the process of seeking out a meeting with Mandela with my strategic advisors, and after much back and forth, we were able to get a brief engagement with him. I was allowed to bring four leaders with me, according to the protocol that his office gave us. I invited Paul Okalik, who was premier of Nunavut at the time; Carole Mills from CAIPAP; Cindy Dickson, a Gwich'in woman representing the Arctic Athabaskan Council; Chief Robert Charlie of the Yukon First Nations; and Larissa Abroutina, vicepresident of RAIPON. My two entrusted advisors, Terry and Stephanie, joined us on the trip, although they were not allowed in the meeting room itself.

The day of the meeting, Paul Okalik and I wore our traditional *silapak*s, ceremonial coats worn for special occasions. When we arrived at the house, we discovered that Mandela's home was a highly secured compound. Every vehicle needed to receive permission at the guard house to enter the driveway. (Our taxi driver thanked us, saying that it was Inuit from Canada who made it possible for him to drive into the house of the man who had freed his people.) At the house, Mandela's assistant oversaw his time down to the minute. Moving through all the security and the handlers, I wondered if Mandela ever felt as if he were still imprisoned in some ways.

Although it lasted only half an hour, the meeting was memorable. As I was head of the delegation, I allowed the other leaders to precede me, sitting before him, one by one, to say what they wanted to say. Mr. Mandela insisted we sit very close to him, as he was hard of hearing. I could also see that he was having difficulty walking, and I sensed his legs were giving him trouble. It was understandable considering his age and what he had gone through during his many years of imprisonment.

What an honor I felt when I finally sat in front of him. He exuded a gentle, kind wisdom, much like that of our Inuit elders. One felt calm and safe in his presence. As usual, my *qallunaaq* look generated some confusion. His first question to me was, "Are you Inuit?" I answered yes and felt his full respect for that fact.

To meet Mandela, to be in his presence, was enough to give me the strength and focus to deal with the issues at hand. However, my team had a strategy in mind for this historic meeting. In light of the challenges of getting a treaty to ban POPs at their source, we felt that a public statement from Mandela supporting our plight would strengthen our bid to secure a strong and binding convention. I had requested this of him by letter before the meeting, connecting the child-centered work of the Nelson Mandela Foundation to our task of ridding our children of toxins. Now I asked again in person. His response was almost apologetic. He told me that all his advisors were away in Cape Town, and without their involvement and expressed approval, he could not, in his words, even dot an *i* or cross a *t* on any public statement. It was disappointing to hear this, but I had one more thing to ask of him before leaving.

In his book *Long Walk to Freedom*, Mandela describes a time, after his release from prison, when he was on his way to Ireland after meeting with Prime Minister Brian Mulroney and addressing the Canadian Parliament. The small jet he was on made a refueling stop in Frobisher Bay, Northwest Territories (now known as Iqaluit, Nunavut), before crossing the Atlantic (although in the book he misidentifies the landing spot as Goose Bay, Labrador). Inuit in the community found out that Mandela's aircraft had landed and were eager to get a glimpse of him if he did, by any chance, get off the plane. Many rushed to the airport, including Anne Hanson, former commissioner for Nunavut, and Ken Harper, a former school-teacher and now a businessman and historian. They stood up against the wire fence. David Serkoak, a teacher, was there along with his daughter. He had brought his traditional Inuit drum. He played in honor of Mandela and what he stood for, hoping that Mandela would hear him. Mandela indeed got off the plane and went for a short walk. He saw and heard the people up against the fence, and asked a Canadian official, "Who are those people?" The answer came back: Eskimos. His immediate response was, "I want to go see them." Someone placed a blanket on his shoulders, and off he went, walking right up to the fence where Inuit of Iqaluit were eagerly waiting for him. He repeated this story to me as I was sitting in front of him. "And there they began to sing and drum dance, expressing support for my life and elation for my release," he said. He added, "Here I thought Eskimos were mainly seal hunters. I didn't know they were so global and knew all about my life." When he told me this story, I thought to myself, *Wow, it's amazing how this came full circle.* Two Inuit leaders, Paul Okalik and I, hearing this story first-hand all the way in Johannesburg in Mandela's home. It was yet another moment of affirmation for me that all things do connect.

I realized this was the opening I'd been waiting for, and relayed to him the story of Elisapee Davidee, an Inuk woman, who, when she heard he had landed, raced to the airport with the "Free Mandela" T-shirt she had worn during his imprisonment. She hoped he would sign it for her. I told him how disappointed she was that she missed him. However, when she heard me on CBC Radio talking about my upcoming meeting with Mandela during the POPs negotiating sessions in his country, she thought, *Here's my second chance.* She asked if I would bring the T-shirt with me for him to sign. We had been given strict instructions to bring any books to be signed by Mandela the day before our actual meeting with him, but I'd taken the risk of bringing the T-shirt with me. I asked him if he would sign it for Elisapee. He glanced at his assistant, who had already informed us that our time was almost over, and said, "I will do it!" And he did. I was so thrilled to be able to bring this symbolic signed T-shirt back to Elisapee. Global Inuit indeed!

When, some years later, I heard that Mandela had died, I once again felt honored and blessed to have had the chance to witness the energy and strength he brought to the fight for human rights. I have no doubt that after meeting him, I carried this energy with me in my own efforts to link climate change to human rights. A heartfelt *nakurmiik*—thank you—Madiba. I am grateful for having met you and heard, in your own words, how meaningful your meeting with Inuit was on that cool summer night in Iqaluit, Nunavut. It means the world to us who live at the top of the earth.

THE FIVE INC SESSIONS were crucial opportunities for Inuit and Aboriginal peoples to draw attention to our plight. But the work that CAIPAP and the ICC team did in this regard was not restricted to the INC meetings. At the first INC session in Montreal, I met Barry Commoner. Barry was a scientist from the United States who had founded the Center for the Biology of Natural Systems at Queens College, City University of New York, to promote research on ecological systems. Barry had worked on dioxins and furans, two POPs that we were finding in our food. These chemicals are produced through the burning of waste, plastics, fuel, tobacco, and wood; through the production of iron and steel; and by electric generators. The health impacts show up in problems with a person's skin, liver, immune system, endocrine system, reproductive system, and nervous system development, and in particular types of cancer.

Barry was focusing on the long-range transport of dioxins—highly toxic compounds produced by cement kilns, copper smelters, medical and municipal waste incinerators, iron smelters, and other industrial processes. Drawing upon the detailed inventory of dioxin releases in the United States and secondary impacts in Canada and Mexico, Barry developed a computer model to calculate not only the levels of dioxins falling upon communities in Nunavut but also the percentage of that total contributed by individually named industrial plants (mass balances and back trajectories). This was remarkable work.

Barry's study found that Nunavut itself could have, at most, contributed 0.32 percent of the total toxins found in Nunavut, while sources outside North America contributed only between 2 and 20 percent. This meant that almost all the dioxins and furans found in the Nunavut environment were a result of North American sources outside of Nunavut. It was a remarkable study that helped us to strengthen our claims and gave us a strong argument to get polluters to clean up their acts.

Using this information, I wrote to the ten or twelve most offending polluters, as identified in Barry's report, including U.S. Steel, Bethlehem Steel, Southwire Company, Lafarge Alpena, and Northern States Power Company. I told them that they were harming Inuit of the Arctic and advised them to reduce their emissions. Some of the companies that responded were dismissive of Barry's methodology, while others said that they didn't know this was happening and would look at how they could better implement their own policies to minimize the damage to our Inuit world. Others didn't respond at all.

While the responses to my letters were mixed, Barry's work, and my use of it, did generate media attention. After the public release of his findings in New York, Barry and I did a number of interviews about this issue. I am forever grateful to Barry for both his research and the encouragement he gave me to carry on.

My work with Barry was only one of the many prongs of our campaign against POPs. Terry was tireless in securing my participation at various events outside of the UNEP meetings. He also worked to connect us with people who he thought might have some influence with those in power. At one point, Terry organized a meeting in New York between me and former undersecretary of the United Nations Maurice Strong. For many years, Strong had been the head of the Rio Earth Summit process that started in 1992 and that would eventually lead to the Kyoto Protocol and other international environmental agreements. He had just returned to New York from China, so Terry and I waited outside his office for him to arrive. When he came into the reception area, he scanned the room. I guess he was expecting to see a dark-looking Inuk woman. Instead, he saw fair-skinned me. "I thought the Inuit were here," he said as he passed the reception desk.

"Yes, that would be me," I said. I wasn't surprised. Over the years I had gotten used to having to clarify my heritage. Maurice Strong caught himself. "Oh, okay," he said, and invited us in.

We went into his office and shared some of the work we'd been doing. Maurice Strong has a long and impressive history in international environmental policy, dating back to the highly successful United Nations Conference on the Human Environment in 1972. In 1992 he acted as chair of the United Nations Conference on Environment and Development, after which he served as the first executive director of UNEP. But what I remember most about the visit was the story he relayed during our meeting. He explained that at seventeen years of age, he first went to the Arctic to work in one of our Nunavut communities for the Hudson's Bay Company. He said that everything he knew about the environment, the Inuit had taught him. It was a remarkable statement, and one that really encouraged me. It was comforting to know that a man of his stature, and in the role that he played, understood what we were up against, and what we were working so hard to protect: the Inuit way of life as it related to the rest of the world.

Maurice Strong's words would continue to resonate with me over the years. During my time working on the POPs issue and in the years to come, I would travel all over the world. I would meet and work with international leaders, influential politicians, and Hollywood celebrities, including people like the Prince of Norway Haakon Magnus, Prince Albert of Monaco, Paul Martin, Jean Chrétien, Bill and Hillary Clinton, Salma Hayek, Brad Pitt, Jake Gyllenhaal, Robert Redford, Ted Turner, and so on. In other words, I had the ears of many powerful people and the media. This did not always endear me to others in Aboriginal communities. Some Aboriginal leaders felt that I was too focused on Inuit in my speeches and presentations. But with the ICC's history of involvement with the POPs work, and with our exposure to toxins being the greatest, I thought it was a good strategy to focus on our experience. (Shirley Adamson, John Burdeck, and Will Mayo, all Aboriginal leaders with a wealth of experiences, supported this point of view.) Some of my fellow Inuit, from ITK, for example, also felt that the spotlight was more on me than it should have been. But I was always aware that none of the attention was actually about me—it was, rather, about my position and my mandate. I knew I had a responsibility to those who'd elected me, and that what people thought of me was secondary. I was, after all, trying to share the collective wisdom of our people. As Inuit we appreciate the wisdom of the land. The land teaches us what we need to survive. It builds character, nurturing judgment, courage, patience, and strength under pressure. And now it was teaching us about the connectedness of all human activity— both harmful and positive—on the planet. Maurice Strong had said it—everything he knew about the environment he had learned from the Inuit. *So had I.* It was the Inuit way of knowing, not just the scientific way, that I tried to share with my audiences. Giving voice to this collective wisdom was always deeply humbling.

The fight against POPs, however, was both a political journey and a personal one for me. I poured everything I had into that campaign. It taught me a great deal about commitment, courage, and perseverance. It was never just a political fight for me. More than anything, I wanted to change the world for the people I cared about. Our people. Future generations. And, of course, my family. What was precious to me was what kept me going. So when tragedy struck, I was especially devastated.

Six days after I arrived home from Nairobi, at six thirty in the morning, I received a phone call from Pamela, my sister's oldest child. Hearing Pamela's voice on the other end set my mind racing. Something had to be terribly wrong for her to be calling me so early. I knew if anything had happened to a family member, it would be Bridget who would call me. In those split seconds, I was sure that something had happened to Bridget's youngest child, Steven, and that she was with him, which was why Pamela was calling. Speaking in a calm tone, my niece told me that her mother had just passed away in the Kuujjuaq hospital. Bridget had had a massive heart attack. She was just forty-eight.

My entire world stopped.

No doubt in shock herself, Pamela gently proceeded to tell me that she had received a call from her mother during the night, as her father was away for meetings in another community. Pamela had driven her mother to the hospital. My sister had thought she had food poisoning—not surprising as the symptoms of heart attack are so different for women than they are for men. Bridget then suffered two heart attacks after being admitted and could not be revived. With great caring, Pamela, knowing how close my sister and I were, said that she didn't want me to hear the news from anyone else.

Sitting in my living room in the early morning of that dreadful day, I held a picture of my beautiful sister in my mind. Her father, Patrick Doyle, had been Irish, and the wavy coppery-red hair of her childhood had deepened to a rich auburn as she got older. Freckles, another inheritance, had dotted her face as a child. While Bridget had never met her father, she had always honored her link to Ireland. She loved the color green and acknowledged St. Patrick's Day in her quiet way. She saved four-leaf clovers and collected several tokens from the Emerald Isle. I felt such deep sadness for her children, Pamela, Tina, and Steven, and for her eight-year-old granddaughter Penina (named after my late aunt Penina), whom she was raising. She also left behind four other grandchildren who would never get to know their gentle and caring grandmother.

Since my grandmother's passing, Bridget had been much more than big sister to me, ensuring that I was okay emotionally well into my adult life. She stood by me every step of the way, in the triumphs and the challenges of my life, from childhood to adulthood, with my work and with our family legacies. She was, in fact, with her gentle demeanor, my protector on many levels. But Bridget was not only a caring sister. She was also a great teacher, working for years at the Kuujjuaq school. She was a loving mother and grandmother, and someone who never wanted or tried to draw drama into her life. As hard as it was to lose her, I find solace in the belief that her passing brought peace for her soul, and it is that belief that has allowed me to carry on without her.

Still in shock for many weeks after this devastating news, I couldn't understand how everyone else carried on with their lives. I felt paralyzed, even though I went through the motions of living. It was as if a part of me had died along with my sister. I flew home for the funeral with other family members who lived in Montreal and with my brother Charlie, who, being a federal senator, was in Ottawa at the same time as I was when we got the news. My mother, in the meantime, had become very ill with pneumonia and was hospitalized, missing the funeral entirely. The shock of losing her special *panik* (daughter) took its toll on her and, with a heart condition herself, likely shortened her own life. My mother was never the same after that, and she died three years later. Losing my mother was also difficult; I felt I was losing not just what I had but also what could have been. Although in life my mother couldn't always be there for me emotionally, the moment she passed away, I felt her presence at every turn.

During the years of negotiating the POPs protocol, and for several years after, I was tested over and over again. Soon after the loss of my sister, I lost my aunt. A year later, my mother, then my young cousin, and later my niece. During this same period, I also almost lost my own child. Sylvia suffered a life-threatening medical complication, and had to be airlifted from Kuujjuaq to Montreal for surgery. She nearly didn't make it, and the days surrounding this emergency were harrowing.

As I poured myself into my work, my spirit was continually tested, as an individual, as a woman, and as a mother. Although these very personal losses shook me at my core, and I thought I would never stop grieving, unexpected openings and changes in perspective helped to renew my inner confidence and deepen my spirit. When you lose loved ones, your perspectives change in profound ways. Family traumas surface in order to allow you to work through them. The shock of it all loosens up everything that needs attention and puts you on a fast track to heal intergenerational wounds. For the next few years, I looked inward, honoring how my lost loved ones had been a gift to me while they were living, and a source of perspective as I grieved their passing. My losses helped me recognize my life's calling and built my character, lending me courage leading up to my future work.

AFTER THESE PUBLIC and personal struggles, the negotiations wrapped up. On May 23, 2001, the delegates adopted the Stockholm Convention on Persistent Organic Pollutants, a legally binding agreement. The negotiations had been tough, and in order to get an agreement in place, environmental advocates had had to make some concessions in their bid to get rid of the "dirty dozen." Those signing the convention would agree to stop the use of only nine of the twelve identified POPs. The agreement had also backed away from the complete elimination of DDT use, stipulating that its use would be limited to the prevention of malaria. The convention also called for the reduction, not complete elimination, of unintentional production of dioxins and furans. But there were also provisions for adding new chemicals to the list over time.

At the signing, the organizers of this historical event had

purposefully seated me near the front so I could have a close view. Our own environment minister, David Anderson, was the first delegate to read, sign, and ratify the agreement. As I watched him and representatives from all the other countries come up to sign the treaty, I was overcome with the feeling that the world was still compassionate and loving, that humanity was still good. I recall smiling throughout the entire proceedings. I was proud that we had been able to achieve all this by engaging in the politics of influence, rather than the politics of protest. We had achieved this by reaching out, by talking *with* industry reps, governments, and other international organizations. We had poured our energies into making presentations at the negotiations and talking with the media. We had relied on our powers of persuasion, on our abilities to get people to want to work with us. So many came together for a common goal: ICC, CAIPAP, environmental organizations, industry representatives, IPEN, and government delegations.

After the signing, a number of members of Environment Canada approached the ICC and CAIPAP team. I think they felt encouraged by the partnership that had been built between the government and our Indigenous organization. Barry Stemshorn, one of the deputy ministers for the environment who was charged with the POPs work, instructed his staff to come to meet us and ask us if there was a way they could further support our work to ensure that the Stockholm Convention would be ratified and implemented by the countries that had signed, and that others would be encouraged to sign as well. "What can we do to build this partnership and foster it?" they asked. It was a rare event, an arm of the Canadian government approaching an Indigenous organization to say, "What can we do next together?" And it was a bright moment when the government of Canada and Indigenous peoples sang from the same song sheet and expressed the wish to do so in the future. This was a partnership that we were going to build on. Indeed, for the next couple of years, we received additional funding from Environment Canada to help us work on these issues.

At the closing gala in Stockholm, I presented John Buccini and Klaus Töpfer with ostrich eggs that I had asked a young Inuk artist and mother, Celina Kalluk, to paint with Arctic scenes. I'd seen similarly painted ostrich eggs in many stores in South Africa, and the gift symbolized, to me, a bridging of North and South. The eggs were well received, and I got a lovely memento in return.

I have a habit of picking up pennies when I see them lying on the ground. I call them my "pennies from heaven" or "angels" and feel that they keep me connected to the source, and reassure me of my path. Without knowing I had this connection to pennies, that same night at the gala, Klaus Töpfer presented me with what he called a "lucky global penny." He had given the coin to John Buccini at the onset of the negotiations. John returned the penny to Klaus that evening, and Klaus passed it on to me. A gift from my angels.

The Stockholm Convention signaled a new era of global management of chemicals and global responsibility for protecting the planet. Although in 2001 the convention was initially signed by over one hundred countries, we had to wait for fifty more countries to officially ratify it (in other words, get approval from their elected members of the government) before the convention could enter into force and the obligations could take effect. This happened on May 17, 2004, ninety days after submission of the fiftieth country, which was France. At the time of this writing, the convention has 152 signatories and 179 countries have ratified it. The United States has not ratified the agreement and remains an observer. There is a lot of work to do to ensure that more countries continue to ratify the convention, and that countries have the capacity to meet the obligations. We also need to move beyond the obligations of the initial convention to ensure that our children live free from toxic chemicals in our environment. To this end, nine new POPs have been added to the list since the original convention. New POPs continue to be identified and are being submitted for listing in the treaty. Negotiations about liability and redress issues continue, and the ICC remains involved in the process.

Through the implementation of the regulations under the agreement and other actions, the presence of many of the original dirty dozen in the Arctic has declined. The lessons learned through the POPs protocol process were also applied to developing a global mercury agreement, which was signed in 2013. Again, the Inuit Circumpolar Council was a key player, with Dr. Eva Kruemmel acting as ICC technical advisor and Duane Smith, an Inuk from the Inuvialuit Settlement Region and the president of ICC Canada, taking political leadership.

NOT LONG AFTER THE SIGNING of the Stockholm Convention, I made an important personal decision. I approached the ICC and said that I wanted to move back to the North, to continue my work from Iqaluit rather than Ottawa. At first, the board of directors was resistant. The offices were in Ottawa, and ICC staff have always lived in the same city. But I was insistent. I was on the road, flying around the world, for many days at a time, month after month. After all my losses, after all my time away, when I came home, I wanted to be *home*. I needed to spend time in the Arctic, living among my fellow Inuit. And it was only a matter of time before my daughter and my grandson would be living in Iqaluit.

At the heart of my need to return to the North was my identity as an Inuk woman. My international work and the time I had spent living in the South had done nothing to weaken that. And interestingly, all the time away had helped me accept some of the things about myself that I had struggled with over the years. While I worked with many people from all around the world, many of those involved with the NGOs, the government agencies, the United Nations, and so on, were of European descent. It began to occur to me that my looking "white" might be, in part, playing a subtle role in my interactions with some of these people. When white people first looked at me, they didn't see "other." (I remembered Maurice Strong, wondering where "the Inuit" were.) They saw someone who looked like them— and who sounded like them. In a world that focuses so much on "othering" one another, it seems my look of "sameness" was one less obstacle in moving Inuit issues forward and helped to alleviate some of the barriers. Over the years, I started to realize that even our "form," whether we like it or not, serves a purpose, and that my looks may have helped start communication a bit quicker, helped lower barriers a bit sooner. I began to sense my role in form and in spirit unite, and my childhood angst about my appearance become more of a non-issue within my soul. To me it finally felt as if my "whiteness" as well as my spirit was serving a higher purpose in my life's work. (As I wrote this book, I wondered if I would be creating misunderstanding with this discussion of my not liking my "whiteness" and how, in my own inner journey, I came to realize it has served its purpose. I share this story because it was a long journey for me to fully accept and honor the form I was born into, looking so white but feeling so Inuk. Then I noticed a Facebook post by a fellow Inuk, who was addressing a post by one of *his* mixed-blood friends. In it, he suggested that this feeling of purpose is felt by others in our communities. He noted that his friend's white-looking appearance had allowed him to be a good mediator between two cultures, helping him to build bridges and infrastructure to support their people. But I acknowledge that mixed blood is not necessary for bridge-building and accomplishment in the wider world. So many of my full-blooded Inuit community have made many similar achievements.)

At the same time, I started to sense at a deeper level that representing my fellow Inuit and the North, fighting to preserve our way of life, fortified my sense of self. I was Inuk— heart, mind, and soul. And I knew deep in my bones that the only place that was truly home was the Arctic.

The move to Iqaluit was a great decision. While I missed Kuujjuaq, Iqaluit was easier to fly in and out of, and it was closer to Greenland, where I frequently had meetings. More important still, my daughter and grandson were moving up there as well. And Iqaluit gave me one of the most extraordinary moments I have ever had with my son, Eric.

A year or so after I moved back North, I received a call at eight at night from Eric. He had flown for Air Inuit for eleven years but was now a jet pilot for a Canadian-owned airline that flies internationally. He had called to say that he was flying that night from Edmonton to London, England, and the flight path would take him directly over Iqaluit. "I'll be overhead at precisely 12:09," he said. "If you go outside then, and look up, you'll see me."

I set my alarm to make sure I didn't miss him. Then, just before 12:09, I put a sweater over my pyjamas and went outside onto the balcony of my house. I looked up to the sky. It was a cool fall night, the inky sky full of stars. At precisely

12:09 the stillness of the northern night was broken. I heard the distant roar of a jet and saw the flashing lights above me. It was my boy. After everything Eric and I had been through, after all his painful struggles in school, there he was, doing what he had always dreamed of—he was literally soaring. I put my hand over my heart and felt my throat get tight, my eyes fill with tears. I was so proud of him. As I went back into my house, memories of Eric's childhood challenges came to mind. From a mother's heartfelt place, deep feelings of not having done enough for him at school and at home had me weeping inconsolably ... until the phone rang. It was the flight service station in Iqaluit asking if I were Eric Cloutier's mother. "Yes," I said, still trying to catch my breath. The man on the other end of the line said, "I am in radio contact with Eric and he wants to know if you saw him fly over." "I did," I said, "and I am still weepy from pride." "Awwwh," he responded as I thanked him for calling. I reflected on that moment for a long time that clear night in Iqaluit. I realized that my sense of pride in my son allowed for some cleansing of the soul, allowed me to let go of the guilt that mothers often experience when they don't feel they did enough to help, guide, and protect their children.

But Iqaluit gave me much more than that special moment. The house that I bought looked out over Frobisher Bay. The water, the landscape, the open skies that I could see from my doorstep were like a balm to me each time I returned home. And being surrounded by family and friends was restorative, a source of strength and energy during the next chapter of my life—my fight for the right to be cold.

6

THE VOICES OF THE HUNTERS

DURING MY YEARS as the ICC Canada president, our work had largely focused on the campaign to eliminate POPs. But this was not the only issue threatening our survival in the North. Climate change, we could see, loomed over the Arctic world even more menacingly than POPs.

Take the village of Shishmaref, Alaska. Evidence shows that people have been living in the island village, hunting and fishing, for at least two thousand years. But today, climate change is threatening to wash away this traditional village. As the sea ice melts, the shoreline is losing its defense against the stormy waves of the Chukchi Sea, which now gnaw away at the shore. Houses that were once far from the water are now falling into the waves. Meanwhile, the permafrost on which the town was built is no longer permanent. As it thaws, houses and roads sink into the soft earth, leaving a topsy-turvy world of dangerous, tilting buildings and suddenly homeless villagers. What could be more devastating than seeing your home swept away by the waves?

Alaskans, like the residents of Shishmaref, had been

witnessing changes to their land and their environment since the 1970s. Alaska is the most vulnerable area in the Arctic in terms of climate change and global warming. But over the course of the eighties and nineties and into the twenty-first century, stories about troubling changes were coming from every corner of the circumpolar North. In my own region of Nunavik, in the community of Salluit, houses had been relocated because of buckling buildings, the result of the permafrost melt.

Everyone in the Arctic communities could see that the ice melted sooner and returned later. And people were always commenting on how different the ice was from the ice we remembered. In the past, we could glance at ice and accurately gauge its depth and its stability. But now, what looked like solid ice might actually be thin or unstable. It was often impossible to say what lay beneath the surface. For a people who spend much of their life moving across snow and ice, this was profoundly troubling—and dangerous. Ronald Brower of Barrow, Alaska, has described his experience with the thinning ice: "One of my sons … fell right through the ice…. I've seen fellow whalers … break through the ice, because it's melting from the bottom, and our snow machines have fallen through."

All my years growing up in Kuujjuaq, I don't recall many stories of hunters having accidents with breaking ice. I remember hearing of one elder and her grandchild who fell through and drowned, but such events were rare. Now we were hearing stories like Ronald's all the time. And when I moved to Iqaluit, I was reminded of the tragic results almost every day. My neighbor, Simon Nattaq, had fallen through the ice on a hunting trip soon after I moved to Nunavut. He pulled himself out of the water and, in soaking wet clothes, dug himself into a snowdrift for insulation. He spent two days that way before he was found. Both his legs were already frozen, so they had to be amputated. When I was home between trips I witnessed his healing from across the street as he slowly but surely learned to get about on his prosthetic legs. As the years passed, I would see this seasoned hunter getting back on his snowmobile or his four-wheeler, going off to the land and sea to hunt. His ability to overcome this great challenge and remain a provider for his family was an inspiring example of the strength and resilience of the Inuk hunter.

But I heard many other stories that showed that even if the ice supported our people and their machines, the thinner ice was making some types of hunting extremely difficult. "You need thick ice for the weight of the whale to bring it up," explains Roy Nageak of Barrow, Alaska. "When it's like three, four feet, especially if somebody got a bigger whale, it's going to keep breaking up…. So, we're trying to catch smaller whales … the smaller the whale, the less [meat] the people get."

And the thin ice wasn't just decreasing the efficiency of the whale hunt—it was making it downright dangerous. Ronald Barrow describes some of the gruesome events he's witnessed: "I believe we now have lost two or three people … because the ice broke, or the rope broke, and when it swung back, it went faster than a bullet. One lady's arm was severed. One, her brain was scattered over the ice because she got hit…. And the third one, her jaw and skull was scattered into hundreds of pieces. But she survived. There's others that have not been so fortunate."

The disappearing sea ice meant that mammals that depended on the ice—seals, walrus, polar bears—were moving farther and farther out, where the sea ice was. Now our hunters had to travel much farther afield, and even then, the harvest wasn't as bountiful. Sometimes hunters would have to take boats out to reach the sea ice, but boat travel through ice-filled water was dangerous, and many hunters found the cost of fuel alone made the hunt far less productive. In Nunavut, people reported that since the ice was becoming so unreliable in the warmer springs, they could no longer travel out to the islands to collect eggs, geese, and seal. By the time they could get to the islands by boat, the eggs were too old to harvest.

Indeed, the loss of sea ice was causing real travel challenges. Hunters complained that with the unreliable sea ice and some areas not freezing over at all, they were often unable to cross the bays and instead had to follow along the shorelines, making their routes much longer and giving them less time to hunt on each trip. Even when the bays were frozen enough to cross, the surface ice was now rougher, crumbly, and less slippery, making travel much slower.

And it wasn't just disappearing sea ice that was creating problems. Many of our communities were reporting that shore ice was shrinking year after year. The shore ice provides protection from coastal erosion caused by waves and water surges, which were increasing in severity because there was more open water and higher winds. Soon Shishmaref was not alone in seeing its shoreline slip into the sea.

On the radio, on TV, and in the community gathering places, people also talked about the changes they saw in the snow. Some winters, there wasn't as much snowfall, and when the snow did come, the warmer temperatures produced a softer, stickier snow. The lack of crisp, dry snow made the running of sleds and snow machines much slower. And the inadequate snow coverage on the ground damaged sled and snowmobile runners, and hurt the dogs' feet. Even walking through the softer snow was more difficult.

Many people were talking about how the deep, dense snow required for building snow houses was extremely hard to find. Lucas Ittulak of Nain, Newfoundland and Labrador, puts it this way: "The snow is not the same anymore. The bottom of the snow is a lot softer than it used to be. It's no good for igloos anymore. [Twenty years ago] we used to be able to stop anywhere we needed a place to sleep just to build an igloo and sleep in that. And nowadays you can't just find good snow anywhere." When you can't find good snow in the Arctic for shelter, something is definitely wrong.

Hunters were having to bring tents with them on their trips. The tents not only were bulky burdens, but also didn't provide adequate insulation in the winter cold or protection from bears the way snow houses did.

Many in our community were also noticing the changes in the animals we depended on. While many of the marine mammals were now much farther north than they used to be, making hunting them difficult, we noted that the animals also seemed less plentiful and less healthy. Breaking ice and storm surges could wipe out litters of seal pups or separate offspring from their mothers. The seals often appeared to have less fat, which meant they sank deeper in the water, making them tougher to catch. Polar bears, hare, and ptarmigan also seemed thinner. Some caribou were thinner, too. The changing melt and freeze cycles often left the tundra covered in a layer of ice instead of soft snow. The caribou couldn't forage for lichen and other vegetation under the ice, the way they have to in the winter. In some areas, however, the longer growing season and increased vegetation meant the caribou were getting fatter. The ocean fish also seemed to be changing. A lot of people commented that the flesh was softer, mushier than it had been, perhaps because of the warmer waters. We heard that there were fewer lemmings in the Arctic, which was troubling, as these are the primary food of our Arctic foxes.

I recall listening to Caleb Pungowiyi of Alaska, at a conference in Seattle, Washington, telling a story of the changes in the Arctic. He explained that, in the spring, the rivers do not "break up" in big chunks anymore. Instead, the ice just melts away bit by bit. As a result, the riverbanks are not scoured by moving chunks of ice, which means that the dead tops of the willows along the riverbank are not clipped off. Without this natural "pruning," new growth of the willow does not happen. Then the caribou, which come along in early summer after the breakup to eat the new growth, go hungry and grow thinner. Once again the changing Arctic weather shows just how interconnected nature is and how humans, plants, and wildlife are all dependent on a stable climate to survive and thrive. (Caleb has since passed away, and we all miss his calm and reflective manner and his wisdom on these matters.)

The changing sea ice was also changing polar bear habits— in dangerous ways. With smaller ice floes moving closer to the land, polar bears were showing up more and more often in our coastal communities.

But perhaps even stranger for us was the fact that new species were arriving. I'll never forget being outdoors in my backyard in Iqaluit and seeing a bird hovering in the air, like a hummingbird. In my many years in Nunavut I had never seen this type of bird. I wondered what it was. I was later told it was a wheatear, which had long been in Nunavik but not in Nunavut. Another afternoon, in a friend's backyard, we saw a robin. It was such a rarity that we were all glued to the window, watching it. I grew up with robins in Nunavik: we used to walk the land to find their nests with their bright-blue eggs, but until recently, they weren't seen this far north.

These were not isolated incidents. Many were seeing robins, pintail ducks, barn owls, and salmon in the High Arctic for the first time. And unfortunately, the longer periods of thaw meant that mosquitoes and blackflies were thriving, and the destructive spruce beetle had moved north in the western part of our homelands to attack our trees.

But perhaps most shocking of all was that the very ground beneath our feet was no longer solid.

Not far below the surface of the Arctic tundra is permafrost, a mixture of gravel, soil, and ice that never melts, even in the summer. At least that's the way it used to be. As our Arctic temperatures have edged upward, the top layers of permafrost have begun to melt (or put another way, the permafrost starts much farther below the soil surface than it used to). As a result, what was once terra firma is now unstable, slumping terrain. This soft, loose earth is vulnerable to erosion. And just like in Shishmaref, houses throughout Nunavut, Nunavik, and the rest of the Far North were shifting, their walls and foundations cracking. More and more communities were watching their roads buckle, their airstrips heave and split, and their rail lines twist and sink. Fissures and breaches were appearing in oil and gas pipelines. In 1994, in Usinsk, Russia, a ruptured pipe resulted in a huge oil spill, making the polluted land unusable. Shifting earth made the prospect of more cracked pipes likely.

The loss of permafrost was also making our traditional way of storing food from the hunt less dependable. Our hunters usually dig cellars deep into the frozen earth to cache meat for months at a time. Many of these caches were becoming unusable as the permafrost disappeared. In 2005 Eugene Brower described his problems with caching: "I've got one ice cellar that's about twelve feet into the permafrost.... Even with the layer of five feet of snow on the bottom ... my game is melting on top.... It's not frozen solid.... They're going to spoil.... I've got another ice cellar that's about twenty-five feet into the ground, and you're starting to feel that in there too."

The melting permafrost was also giving our hunters more travel headaches. Normally dry tundra was turning marshy and muddy, and many drivers complained that their four-wheelers were getting stuck as they tried to get to and from their hunting grounds.

By contrast, the disappearing permafrost meant that in some areas, the ground was more porous, and water was draining away from the surface. In this way, whole lakes were disappearing into the earth and rivers were drying to a trickle. In some cases, rivers became so shallow that fish could no longer reach their spawning grounds.

The early thaws and later freezes were wreaking havoc on the traditional hunting seasons. The later freezes translated into a later winter hunt. "By the time they get out there, sometimes the game's gone because they go with the cycle," explains Eugene Brower. But the early and often sudden spring thaws were also creating all sorts of problems for hunters. Spring hunting was much trickier, as the snow could suddenly disappear—taking with it our hunters' travel surface. David Haogak of Sachs Harbour, Northwest Territories, has described how early melting and increased spring rains mean that the water levels of the rivers rise suddenly, making it impossible for hunters to cross back on snow machines to return home. As a result, he says, "No one goes very far now, whereas ... even a few years ago, we used to go forty or fifty miles. Now we don't even go twenty. It's just not worth the risk."

Increased rains and early spring melting were also causing huge runoffs and floods that eroded land, and made swollen streams and rivers uncrossable by caribou herds. Our hunters were discovering, to their dismay, that the herd migratory routes were changing as a result—and they no longer could predict where the game would be.

As if all this wasn't enough, elders were reporting that they couldn't depend on the skies either. One of our traditional skills was learning to look at cloud formations and judge wind speeds to predict the weather. But wind patterns and cloud formations have altered—elders can no longer tell by looking up whether a storm is rolling in. This unpredictable weather means that hunters have been caught off guard by storms that arrive unexpectedly—sometimes with deadly results.

Warmer temperatures and strong winds also affected some of our traditional practices. Drying caribou hides, for example, was more difficult as the hides dried too quickly, making them prone to cracking and tearing.

And finally, the thinning ozone and the increased ultraviolet (UV) radiation at the poles were affecting our people's health. Reports of skin cancer, cataracts, and rashes were on the rise. For the first time in our lives, many of us Inuit had to wear sunscreen and sunglasses when outdoors. And some found the warmer summer temperatures uncomfortable.

Those of us who lived in the Arctic heard these stories every day, and we knew that others needed to hear about the Inuit experience of climate change. After the Stockholm Convention had concluded, the ICC turned its focus to the catastrophic events on our doorstep and began a campaign to add our voice to the international work on climate change.

Of course, a global effort to draw attention to the issue of climate change had been under way for many years, including work done through the United Nations and the United Nations Framework Convention on Climate Change (UNFCCC).

The UNFCCC came out of the ground-breaking Rio

Earth Summit in 1992. Several environmental issues were hammered out in Rio, with 172 countries represented, including 108 that sent their heads of state (famously, Al Gore represented the United States), and 2,400 representatives of NGOs. The ICC was there, represented by then international president Mary Simon and the executive members of the ICC from the national offices. It was a truly sprawling and incredibly ambitious negotiation, covering topics as wide-ranging as water conservation, public transport, toxic waste, and, most critically, greenhouse gas emissions. Though no binding targets were established, it was the commitments set in Rio that led to the Kyoto Protocol five years later. Kyoto has been criticized for being too ambitious and not ambitious enough, both for being too punitive on developing nations and for being too lenient, and for manifestly failing to halt greenhouse gas emissions; yet the fact remains that for the first time, the planet figured out it had a problem and got together to solve it. This was also the first time that the international community set binding targets for itself. (Though, notably, the United States failed to ratify Kyoto, and Canada withdrew from it.)

Every year, participating countries gather to fine-tune the UNFCCC's aims and compliance at a Conference of the Parties (COP). The parties meet annually to negotiate more rigorous measures—since Kyoto, new targets have come out of meetings in Bali, Indonesia; Copenhagen, Denmark; Cancún, Mexico; Durban, South Africa; and Doha, Qatar. And though countries send delegations to negotiate, they are far from being the only participants in the talks. NGOs, private-sector representatives, and Indigenous groups also attend the talks to lobby and put political pressure on the national delegations to seek a binding agreement. These groups also work to raise awareness about the talks at home through the media and side events.

This process may not be ideal, but it's necessary. I say it's less than ideal because it is a political process—it's meant to reach compromise to satisfy as many stakeholders as possible, probably including parties who want no limits on greenhouse gas emissions at all. That is to say, like many, I feel that targets should be set by people who understand the science, and not be watered down by those who benefit from the status quo. Still, the world must gather to figure things out, even if some of the voices at the table aren't helpful. For all its shortcomings, the UNFCCC is the only international body that brings countries together to deal with climate change.

If this UN process was the way to get heard, then the ICC and circumpolar peoples needed to be there. Indeed, the stated objectives of the UNFCCC contain assertions that we are uniquely qualified to support. The goal of the UNFCCC is the "stabilization of greenhouse gas emissions in the atmosphere at a level that would prevent dangerous anthropogenic interference with the climate system." That one word—*dangerous*—brought the whole UN process into focus for us. The Arctic could provide *ample* evidence that climate change was indeed dangerous. We needed to draw the world's attention to that evidence.

We were learning that the poles are warming more quickly—twice as fast, actually—than other parts of the globe. For a long time, scientists assumed that this was a result of melting Arctic ice. White snow reflects the sun's rays, while open water absorbs them. It seemed obvious that the more the poles melted, the more susceptible they would be to further melting. And they are, though that turned out not to be the whole story. Scientists began to notice that climate variability was greatest not in the summer, when the sun was strongest, but in the winter. They concluded that it's not the sun that is conveying all that energy to the poles. It's the planetary weather systems—the same patterns that shuttle POPs away from warm climates toward the cold. Just one more way in which everything is connected.

What that means, of course, is that whatever happens in the poles will eventually happen everywhere else.

For those of us who called the Arctic home, there could be no doubt that the climate was changing. And certainly by the start of the twenty-first century, there was overwhelming consensus among scientists that the world was experiencing grim climate consequences as a result of environmental damage. The Intergovernmental Panel on Climate Change (IPCC) released its third assessment report in 2001. The IPCC is a UN body that assesses peer-reviewed science and draws conclusions about the way the climate is changing. Thousands of scientists from around the world, working on a voluntary basis, contribute to each report. The third assessment report announced that the 1990s had been the warmest decade ever reported, and projected that in the twenty-first century, the planet would see the fastest rate of warming in the past ten thousand years (that is, the span of planetary history during which human civilization has come into being). The third assessment was more dire than the second (just as the fourth was scarier still, and as the fifth is). So it was safe to say that the world's scientists had made up their minds. And yet, anyone who listened to talk radio or to the speeches of conservative politicians might have gotten the impression that no one was really sure what was going on. Prime Minister Harper himself called the science "tentative" and "contradictory," and suggested that the damning report written by the world's top researchers was no more than a "hypothesis." In his famous words, he believed "the jury is still out."

But no one in the Arctic could possibly have believed the jury was still out. Certainly no one in Shishmaref, watching neighbors' homes disappearing into the ocean, or hunters breaking through the ice into the frigid Arctic waters.

And, of course, we weren't the only people with this fight on our hands. Citizens of the South Pacific islands had been experiencing similar destruction of their homeland as their islands were slowly sinking with the rise in sea levels. When their homes disappear, a generation of refugees will have to tell the next about a country that exists only below the surface of the water. But there are countless other implications for the rest of the world. Farmland will dry up as mountain snowpacks melt, leaving riverbeds parched. Some crops, like rice, will not germinate in higher temperatures. Whole regions will become uninhabitable as rainfall patterns shift. Disease will spread. Species will go extinct.

While the issues of POPs and climate change are intertwined, more so in the Arctic than anywhere else, it was impossible in the late nineties for ICC International, with its limited resources, to tackle both at the same time. The timing of the POPs treaty negotiations and the UN climate change meetings often overlapped, and the respective events usually happened in different parts of the world. We did endeavor, however, to send some representation to the UNFCCC. Bernie Funston, a consultant well versed on northern issues, attended one UNFCCC meeting on behalf of ICC, observing the process and reporting back to us about what was happening.

Rosemarie Kuptana, a former president of ITK and ICC International, originally from Sachs Harbour of the Inuvialuit region, attended the meeting in The Hague, where she presented a film produced by the International Institute for Sustainable Development (IISD), a Winnipeg-based organization. The film,

Sila Alangotok: Inuit Observations on Climate Change, looked at how climate change has affected the traditional and cultural activities of people around Sachs Harbour. In this way, these negotiations heard from Inuit voices early on.

In fact, it was the Inuvialuit who started the Inuit involvement with the UNFCCC. John Keogak and Frank Kudlak, both also of Sachs Harbour, often traveled internationally to climate change meetings to represent the Joint Secretariat of the Inuvialuit Game Council. When the ICC did become involved in the larger climate change movement, and I would see them at these events, I always felt supported knowing they were first and foremost hunters who lived the realities of change.

But while the ICC was pushing forward with the Stockholm Convention on POPs, the Arctic Council was making significant contributions to climate change study. At its October 2000 meeting in Barrow, Alaska, the council's eight ministers representing the member countries signed the Barrow Declaration, which, among other things, announced the launch of the Arctic Climate Impact Assessment (ACIA). A co-project of the council and the International Arctic Science Committee (based at the University of Alaska Fairbanks), and funded by the U.S. National Science Foundation and the National Oceanic and Atmospheric Administration, the ACIA would create three reports: a scientific assessment, a summary or overview volume, and a policy report. These reports would make up a comprehensive study that evaluated and synthesized knowledge on climate variability, climate change, and increased ultraviolet radiation. Focusing on environmental, human health, social, cultural, and economic impacts and consequences of climate change, the assessment was also meant to support a policy-making process and the work of the Intergovernmental

Panel on Climate Change. In other words, the ACIA wanted to collect bulletproof science and make sure that the science led to action.

With the POPs work, there had been a lot of scientific data that we could use to shape international policy and make our interventions stronger. In the early days of our climate change work, the circumpolar communities and organizations didn't have as much comprehensive data focusing on the Arctic. The ACIA promised to be a crucial step toward filling that void, and with the Stockholm Convention concluded, the ICC would be increasingly involved in the project.

DURING THE EARLY YEARS of the ACIA work, a number of processes were happening in tandem. While the research assessment was under way, the politics of the policy work was playing out, and the communication and outreach strategy was being put into place.

We wanted a seat at the table for both the policy work and the science assessment. There was no doubt in my mind that we had something crucial to offer. It came down to what we mean when we talk about science. Science is a body of knowledge, and a way of knowing based on rigorous observation. By this definition, the hunters who criss-cross the ice and snow and embody centuries of observation are scientists. When they describe what is happening to their landscape, the world needs to listen.

The ACIA Secretariat chair was Bob Corell of Harvard University and the World Meteorological Institute. He oversaw more than three hundred scientists from fifteen countries, as well as six Indigenous peoples' organizations, referred to as the "Permanent Participants" of the ACIA process: the Aleut

International Association, the Arctic Athabaskan Council, the Gwich'in Council International, the ICC, RAIPON, and the Saami Council. The ACIA research efforts drew upon both science and Inuit traditional knowledge; indeed, many researchers worked directly with our Inuit hunters and elders.

This was a welcome approach. In the past, some scientists had rejected the observations of our Inuit population as anecdotal and unreliable. Yet it was becoming increasingly clear to our communities, as well as to much of the scientific world, that our hunters—in fact, any of us who survived on the land—were scientists themselves. Our hunters and elders were carefully observing nature. They were finely attuned to small changes. They knew what to look for. They were experts because they had to be: their daily survival on the land and ice depended on it. As Maurice Strong had observed, if you wanted to learn about the environment, there were no better teachers than Inuit. And if you wanted to learn something about climate change, there was no better expert to ask than an Inuk hunter. Indeed, since 1998, the Alaska Native Science Commission, headed by Patricia Cochran, had been conducting traditional knowledge projects, which recorded the observations of Inuit hunters and elders. The ACIA researchers followed suit.

Over the four years of collecting data for the ACIA report, the scientific and traditional knowledge teams assembled a sobering collection of evidence. Their assessment reinforced the fact that climate change was occurring in the Arctic twice as fast as it was in the rest of the globe. Among the other conclusions, the climate researchers projected a temperature rise in the Arctic of five to seven degrees Celsius over the next century. The scientific reports would also confirm what the Inuit populations of the North had been observing: early thaws and late freezes, melting sea and shore ice, increased precipitation, reduced snow cover, disappearing permafrost, rising sea levels, coastal erosion, flooding, suffering animal populations, and the arrival of new plant, animal, and insect species. And while the scientific assessment noted that increased vegetation in the North may also increase carbon uptake, it added that the loss of light-reflecting snow cover and ice will outweigh this, producing further warming.

Many of the predictions in the scientific report fell into this good news/bad news vein. Arctic marine fisheries may become more productive, but freshwater fishing—which supports much of the local population—will likely become increasingly difficult as inland lakes drain or disappear. Marine transport through Arctic waters and access to the Arctic's natural resources may improve, but shifting sea ice will also make water travel hazardous.

But many of the forecasts were simply dire: the thawing ground will disrupt transportation, building safety, and natural ecosystems. Increasing exposure to storms will continue to threaten coastal communities and ecosystems. A number of bird and mammal species are likely to face extinction. Increased UV exposure will have damaging effects on Arctic peoples and on amphibian and fish populations.

The scientific work done in the ACIA provided much-needed support for Indigenous peoples' observations about the changing Arctic. But in terms of the policy document, as opposed to the research assessment, it became clear that the Permanent Participants, our Indigenous peoples' organizations, had work ahead of us if the ACIA was going to include our perspective.

We had wanted the ACIA to include traditional knowledge as part of its research. But we also wanted it to address the health and cultural impacts that climate change was having on Indigenous peoples in the circumpolar North. Changes to our hunting grounds, the hunting season, and ice-dependent hunting techniques meant that our hunting skills were already being badly weakened. And this wasn't the only part of our culture that was disappearing before our eyes.

The difficulties with hunting meant that our people had less country food to eat and were becoming more reliant on southern food—a nutritional and economic challenge, but also a spiritual loss. Unsettled and unusual weather patterns meant that elders were unable to teach the next generation how to predict coming storms. The changing appearance and condition of shore and sea ice, snowdrifts, and the landscape meant that navigational skills were no longer as effective. Many hunters told us that with the absence of dense snow, they couldn't teach young people how to build snow houses. Heather Angnatok of Nain, Newfoundland and Labrador, would later tell us, "We've had incidents where young people have perished in the wintertime. Because they didn't, perhaps, know how to make a shelter, even a temporary one, or a small one." Even the teaching of winter hunting skills to the younger generation was a real challenge with the shorter cold season. Roy Nageak has described it this way: "You need to learn about the weather, the currents, the sea and the ice.... If they're not out there hunting, and the ice is not there, then they're not learning what they need to learn.... The experience is not there." It was becoming increasingly difficult for us to pass on our traditional knowledge, survival skills, and cultural richness to our children.

To Arctic Indigenous peoples, therefore, climate change is emphatically a cultural issue.

As work started on the ACIA, the Aboriginal peoples' organizations pressed the Arctic Council to make Terry Fenge a member of the policy-drafting committee. We trusted his analytical and writing skills, and his ability to see both the detail and the larger picture of the work. We felt he would accurately represent the views and the values of the Aboriginal communities in the North in these working subgroups. We didn't give Terry or anyone else the political mandate to represent our cases, but instead used their expertise to further our cause. With Terry's help, and the work of many others, we were able to get the ACIA to address the health and cultural impacts of Arctic climate change on Indigenous peoples in the circumpolar North. We also asked that the assessment conclude with policy recommendations for preventing such impacts and that traditional knowledge be woven into as many chapters of the assessment as possible.

The ICC, along with the other Permanent Participants involved in the ACIA discussions, submitted a statement that we wanted included at the beginning of the science assessment and the summary volumes. It noted that we Indigenous peoples of the North were part of the environment—we had survived in harsh conditions for thousands of years by listening to the land's cadence and adjusting to its rhythms. It emphasized that climate change was an issue of cultural survival. It also noted that climate change, by threatening the animals we depend on, threatened our very existence, and that the speed of environmental change in the Arctic meant that there was little time for our communities to adapt. The statement called on all Arctic states to inject our Arctic perspectives into the global debate on climate change and to assist northern Indigenous peoples to bring their views and recommendations to international institutions that were addressing the effects of climate change. But we were also asking for a number of things beyond the inclusion of traditional knowledge.

We addressed the Arctic states themselves and asked them to set an example for the rest of the world by adopting and implementing strategies to reduce emissions of greenhouse gases and carbon sinks. We also recommended that the Arctic states commit energy and resources to helping Indigenous people adapt to the challenges of a changing environment. The Arctic states, we suggested, needed to equip Aboriginal communities with information and budgets, and they needed to acknowledge Aboriginal authority to make decisions to protect and promote their way of life.

While the council at first rejected our statement because they felt it was too political, eventually, in 2003, they added it to their policy recommendations report. This was a great accomplishment for us, as it kept the human side of the story in focus.

The policy-drafting committee, which included representatives of the eight Arctic states and the six Permanent Participants, met several times over the following two years. In 2003, two years after the assessment work began, they finished a short but comprehensive report (the London draft), which addressed mitigation, adaptation, research, observations, monitoring, and modeling, as well as communications and education. It would be used to draft the final report.

While the policy work was under way, Bob Corell and I, as part of the ACIA communications and outreach strategy, traveled to various global climate change events. Bob, one of the best science communicators I've ever known, would present the research data. I would tell the human story. Together, we spoke at a UN forum in Nairobi, Kenya; an Arctic Council side event at the World Summit on Sustainable Development in Johannesburg, South Africa (2002); and a conference in Washington, D.C.

I began most of my presentations during those years by evoking the image of an Inuk sentinel. As the ICC increased our involvement in international climate change work, we decided to embrace the idea that our hunters, our people, were sentinels, positioned at the top of the world, watching for danger but also sounding the alarm, warning others of disasters on the way. Whatever was happening in our Inuit homelands of all four countries was about to happen everywhere. Our homeland—the Arctic—is the health barometer for the planet, and as such, we Inuit have a significant role to play globally. The Inuk sentinel was the human face of climate change, but also a figure of traditional knowledge. The image represented our fight to depoliticize the climate change discussion, to have the health of the planet and its people, rather than national interests, determine target emissions for greenhouse gases.

My team was pleased with the work of the policy-drafting committee, and we continued to press for more inclusion of our perspective with other climate change institutions. As we had been successful in highlighting the Arctic and its people in the POPs treaty, we asked that this be the model for an amendment to the United Nations Framework Convention on Climate Change. The UNFCCC singled out various areas of the globe affected by climate change, but not the Arctic. We felt this was a huge omission. We asked that the UNFCCC documents include references to the significant impacts of climate change in the Arctic and on the region's Indigenous peoples. Surprisingly, the UNFCCC did not respond to our request and we continued to push for this inclusion in the years that followed.

In the fall of 2003, the ACIA process hit a bump. The American delegate to the Arctic Council announced that he was under instructions to table a one-pager headed "U.S. Statement on Policy Document." There was no date on the paper. It was not signed, nor was there letterhead or a logo to identify its source. The document stated that a fundamental flaw existed in the policy-drafting process. The ACIA policy documents, it claimed, should be developed only after the governments had an opportunity to consider the scientific report and the synthesis document on which the policy document was based. The member countries could draw their own conclusions. This was starkly at odds with the political declaration made when the foreign affairs ministers had approved the ACIA work plan in Barrow. The point of the project was not only to collect data but also to improve policy. For us, the science was a tool to forge a policy that would save the Arctic and the planet. Science on its own would never stop the ice from melting.

We sensed that this roadblock was being thrown up because the timing of the assessment, which was to be presented to the national governments of the Arctic Council in October 2004, would interfere with the U.S. presidential election campaign. George W. Bush's Republican administration had been adamant that they would not change economic or environmental policies if American jobs were going to be lost in the process. They no doubt feared that an announcement that called for reduced greenhouse gases might have a negative impact on their voters during Bush's bid for re-election. It seemed as if the U.S. delegation was working to keep the ACIA from being released until after the 2004 election.

The move was troubling, especially because the United States provided significant financial support to the ACIA, and therefore held considerable sway over the process. Many of us in the Indigenous peoples' groups felt betrayed by this turn of events. I found myself wondering what the Canadian Senior Arctic Officials (SAOs) had been doing in the closed-door sessions that excluded the Permanent Participants. Had they been supporting our interests and resisting this new American initiative? And we worried that these last-minute changes might delay the completion of the scientific report itself.

The ICC felt that we had to push back hard to prevent American politicking from minimizing the potential influence of the ACIA, reducing it to an academic assessment that would end up on the shelves, collecting dust. The science and traditional knowledge contained in the other two reports needed to be accompanied by a proposal for action if the entire assessment was to be effective—and if the Arctic communities were to have any hope that their world and their culture would be protected.

As the international chair of the ICC, I attended all the high-level Arctic Council meetings (the ones at which the ministers of the eight countries were present), but not all the smaller Arctic Council meetings. However, in response to this move by the American delegation, I made one or two unscheduled trips to Iceland, where the Arctic Council meetings were being held.

I had often found the international diplomatic forums and the global travel taxing. The progress at these meetings can often be painstakingly slow, and coming to a consensus is never easy. Frequently, I would arrive tired and jet-lagged and would wonder if my attendance was really worth it. But this new setback drove me forward, especially when it appeared that the chair of the Arctic Council might be bowing to American pressure. I was one of the few women leaders at these forums, and as an Aboriginal one, I was an even rarer entity. This crisis reminded me once again that I represented voices that often went unheard. My maternal instincts kicked in. And my desire to protect the Arctic and our culture, along with plenty of good strategic advice from my colleagues, gave me the courage and perseverance to increase my participation in the council meetings. In my presentations to and dealings with the Senior Arctic Officials within the Arctic Council, I reminded them of their commitment to the original work plan of the ACIA and the importance of the stand-alone policy document.

Of course, I wasn't alone in these efforts. In October 2003, Chuck Greene, the president of ICC Alaska, wrote a letter to the American SAOs. ICC International had felt that a letter from the Alaska office—in other words, an American Inuit organization reaching out to its own U.S. government—would have a stronger impact than any appeal from ICC International.

In his letter, he declared his concern about the new American position on the ACIA policy document, which he thought was going to be tabled in Iceland. Chuck Greene reminded the SAOs that the work plan of the ACIA, as outlined in the 2000 Barrow Declaration, and which the U.S. State Department had played a key role in developing, required that all three reports—the science assessment, the overview report, and a policy document—be presented to the ministers at the same time. And he noted that by delaying the policy report, the United States would be opening itself up to intense criticism.

While Greene's letter served to remind the Senior Arctic Officials of the importance of the ACIA's policy work to its American Arctic constituents, I was able to address the U.S. government in a different way.

On September 9, 2004, while I was sitting at my desk in my Iqaluit living room, gazing out at the open water of Frobisher Bay, a message from Senator John McCain arrived in my email inbox. It was an invitation to speak to the U.S. Senate

Committee on Commerce, Science, and Transportation at the hearing on climate change on September 15.

This wasn't a complete surprise. My political advisors, Terry Fenge and Paul Crowley, a lawyer who'd been the executive director of the Nunavut Social Development Council before joining our team the previous year, had been working with Bob Corell on this invitation for some time, and I had submitted written testimony to the committee previously. But it was difficult at that time to get our voices heard in the United States, so the opportunity to stand before the Senate committee with Bob Corell and tell them about the plight of the Arctic was an important one.

Terry and Paul had drafted a speech for me to present. The science and technical discussions in the draft were solid, but my gut told me that this shouldn't be a cerebral exercise.

Over the years, I've been asked by a number of people how I felt talking with John McCain and dealing with George W. Bush's administration. "Aren't they the enemy?" they'll sometimes say. I always tell them no, I don't think that way. I've never found an adversarial approach very helpful. Instead, I've tried to put my faith in people's innate concern for others. I reason that if I can connect with them on a personal level, if I can appeal to their heart with solid facts to back up my story, I may be able to change their attitudes and opinions. And the only way I know how to do this is to speak from *my* heart.

As I worked on my draft, I felt a sense of "aliveness." I always know I'm on the right track when that feeling hits my core. It gives me confidence in my role in the larger scheme of things, and it certainly puts to rest that nervous high school student who could barely get through a five-minute speech. In fact, as I walked into the Russell Senate Office Building with Bob Corell and my team, I felt remarkably calm and safe. Wearing my sealskin-trimmed black suit, I knew I belonged there. I was representing my fellow Inuit in all four countries, including the American Inuit of Alaska, and I was addressing an American senator who had proved his concern about climate change by chairing this meeting. I was among friends.

After introducing myself, I laid out the scope of the problem that had brought me to Washington. I described the sinking roads and buildings, the swarms of mosquitoes and blackflies, the invasive plant and animal species, and the melting ice. I invoked the latest scientific warnings: that marine species dependent on sea ice, including polar bears, seals, walrus, and some marine birds, are likely to decline, some even facing extinction, and that warming is likely to disrupt or even destroy our Inuit hunting and food-sharing culture. I acknowledged that people don't damage the earth out of malice. We all want to do what's right—it's just not always easy to understand what that is. If I wanted John McCain to understand us, I needed to appeal to his worthy intentions.

Then I took Senator McCain and his committee back to my mother's childhood. I told him of the arrival of the American military in Kuujjuaq in the 1940s, and how the airstrip they built and the economic activity they created saved many in our Inuit community from starvation. I repeated my mother's words: "We wouldn't have pulled through if it were not for the Americans." After showing Senator McCain what his country had already done for us, how important the role of the Americans has been in the history of my birthplace, I appealed for their help again. I asked Senator McCain to write to Secretary of State Colin Powell and request that the State Department change its position on the ACIA policy recommendations report.

The hearings were held in a large room, and those of us presenting sat at a long table positioned in front of a raised dais. Seated at the U-shaped desk above us were the committee members. When I got to the part of my speech about the Americans' arrival in Kuujjuaq, I looked up at Senator McCain. I saw his body relax. He sat back in his chair, and a half-smile crossed his face. I had clearly moved him. It wasn't until much later that I realized my words had touched his core because I was speaking to a veteran of the U.S. military.

But it wasn't just Senator McCain who seemed moved to understand our plight. Immediately after I delivered my testimony, a woman came up to me almost in tears and gave me a hug. She explained that she was one of Senator McCain's staffers and had been in the back room listening to my testimony on a small monitor. She told me that she was so touched by what I'd said that she felt compelled to come out to the hearing room and tell me. I couldn't help feeling that small act was a moment of universal maternal energy, a coming together to protect the future for our children. I left that day feeling more hopeful than when I'd gone in.

Several weeks later, we heard that the Senate committee had indeed sent a letter to Colin Powell requesting that he reconsider the State Department's suggested revisions to the ACIA process. But if anyone thought that we would have a smooth ride after our Senate appearance, they would have been wrong.

Shortly after the trip to Washington, all members of the Arctic Council received a number of emails from the Icelandic Arctic Council chair, expressing "surprise and disappointment" that information about the council's informal discussions of the ACIA process was being used to wage an open campaign. He was dismayed that what had been discussed in closed-door meetings had been made public. He demanded "a halt to all such disclosure" and asked us to "show the necessary regard for our mutual concerns." Terry forwarded these emails to me, saying, "This is directed at us, I think."

The Arctic Council is a consensus-based, high-level government forum. While the group is composed of the SAOs from its eight member countries and six Indigenous peoples' groups (the Permanent Participants), the latter do not have any voting rights. The emails seemed to suggest that the Permanent Participants also did not have the right to publicly disagree with anything that the SAOs were doing, even if we felt that some of them were moving away from the spirit of the 2000 Barrow Declaration. And yet in connection with the ACIA policy recommendations, we felt that we were standing up for the principles of the Arctic Council itself, while the council was more concerned with avoiding ruffling the feathers of any of its member countries. I suppose the chair's position shouldn't have surprised us. Many of the member countries had strong business, political, and economic ties with each other, and with outside organizations like NATO. Naturally, the SAOs would not want to damage relationships that benefited their countries. Certainly, as the Arctic Council chair, the Icelandic SAO didn't want any public conflicts within the organization on his watch. But we weren't there to protect others' business or political interests. We were there to save the Arctic.

The Arctic Council chair's warnings were followed by another slap on the wrist. The Permanent Participants were told that we would no longer be allowed to send two representatives from each Indigenous organization to the ACIA drafting meetings. This would, in effect, mean that the presidents from each group would go to the sessions, but they wouldn't be accompanied by their technical or political advisors. We suspected that the Arctic Council felt that our advisors were pushing us to challenge the ACIA wording and the process itself, as if we didn't possess our own considerable political and strategic sense. I didn't take the insult lightly.

I wrote to the chair and told him we were not prepared to see the Inuit voice marginalized, and made this same observation at the final session of the ACIA conference in Reykjavik, Iceland, in early November 2004. In my presentation to that conference, I reminded the council of the importance of having a stand-alone policy recommendations document, rather than burying the recommendations for action in the general report that the SAOs provide to their ministers every two years. And I assured the conference that I had said nothing to the Senate committee about the policy recommendations that were still under negotiation. But I defended my actions in Washington and pointed out that Inuit and Indigenous peoples, who were most at risk as a result of climate change, were being criticized for espousing the very principles that the SAOs should themselves be defending.

In my testimony, I also raised questions that I had been asking in letters, communiqués, and other talks for several months. Should only one's country's power be able to produce this kind of response to the ACIA? Did the states fully appreciate that the ACIA was about lives, cultures, and the future of peoples, and that these realities had to inform the policy recommendations that the ministers were given to consider? Or was all this just business as usual for the departments of foreign affairs? I asked them to keep in mind that 35 percent of greenhouse gas emissions come from the eight Arctic Council states.

My focus was the political pressures being put on the ACIA, but of course all of these questions reflected the broader problems that afflicted global environmental protection—the focus on saving jobs at the expense of environmental safety, sacrificing the future to meet today's political demands, and buying the right to pollute.

I also tried to come to the defense of the scientists. Government officials at the council meetings had been making comments about how scientists should leave policy and recommendations to the "policy experts." Of course, this attitude was not isolated to these particular officials or this ACIA process. Scientists around the world were being ignored or muzzled when their research presented challenging truths to governments, industries, or businesses. I reminded the group that it was important to respect the work done by both Western scientists and Indigenous elders in compiling the data in the ACIA. Their co-operation was an exceptional model for how to incorporate both of these ways of knowing into a policy document. Actually, it was groundbreaking: this marriage of Western science and traditional knowledge was at play in almost every chapter of the assessment. I noted that it wasn't helpful for governments to dehumanize scientists as a means of dismissing what they said.

One ally, a member of the U.S. Arctic Research Commission, later observed to me that "the State Department went ballistic," after my Senate testimony prompted Senator McCain and two other senators to write that letter to Colin Powell. Indeed, several months after my testimony, I was challenged at a talk I was giving at the Carnegie Institution in Washington, D.C., in which I described the difficulties that the Indigenous peoples' groups had been experiencing with the politics of the ACIA process. One U.S. State Department official stood up after I'd finished and dismissed the seriousness of the U.S. demand to shelve the policy paper.

I stood my ground and refused to back down. The incident made me realize that despite everything I'd been through, both personally and professionally, I had no desire to stop fighting for our people. The impact of climate change on my grandson's future was much more frightening than anything the American administration of the day could ever do to me. I was in for the long haul.

Months down the road, I arrived at an ACIA reception in Iceland. The ACIA work had just been completed and the council had finally come to a consensus. It seemed that Senator McCain's letter had been effective in adding further pressure to the U.S. delegation, which was already being pressured by the other Arctic Council countries: the policy recommendations had been included as a stand-alone document in the ACIA report. Upon entering the reception room, I found myself being hugged from behind. The person held on for a rather long time, as I struggled to guess who it might be. When the person let go, however, I was shocked to see it was the State Department official who had challenged me at that Carnegie Institution event and who throughout our work had strongly "defended" her government's position. "We did it!" she said, referring to the documents that had finally been accepted by the SAOs. I remember thinking, *Indeed we did, despite your tactics.*

The ACIA was presented at the Fourth Arctic Council Ministerial Meeting in Reykjavik on November 24, 2004. It included both the science, including traditional knowledge, and the stand-alone policy recommendations. After four years of intensive work, the scientific assessment was excellent. In fact, at that time, the ACIA was the most comprehensive and detailed regional assessment of climate change in the world. Its key findings, as mentioned earlier, included that the annual

Arctic temperature was rising at double the rate of the rest of the planet and that increased global CO_2 and greenhouse gas concentrations would contribute to a rise of 4 to 7 degrees Celsius in the Arctic over the next one hundred years. The scientific report also provided evidence of widespread melting of glaciers and sea ice, a shortening winter, decreased snow cover, and increased weather fluctuations and precipitation. The many global implications of the warming Arctic were outlined: acceleration of global temperature increases due to the loss of the reflective ice and snow, rising sea levels, and negative effects on global biodiversity as Arctic breeding grounds and migratory routes were disrupted. And, of course, the report offered significant scientific evidence that the Arctic flora and fauna would continue to suffer.

While some of these facts are now well known from Al Gore's film, *An Inconvenient Truth,* and the 2006 Stern Review (a study commissioned by the British Treasury and led by British economist and academic Nicholas Stern), at the time of the ACIA report, this information was new—and shocking.

The policy document, which we'd worked so hard to include, was more than many expected but less than we'd hoped for initially. A modest breakthrough, in other words. It did include recommendations for future research and long-term climate monitoring. It stressed the importance of identifying and evaluating potential measures to mitigate and adapt to changes in the Arctic environment. It addressed the need for improved communication and education about Arctic climate change—for those who live in the Arctic *and* for those who develop local, national, and international environmental policy. What was truly alarming was the cultural dimension of the report. Learning of changes to the climate a few decades down the road is an abstraction.

We know it's bad, but we can't *feel* it's bad. Warnings of food shortages and extinctions of wildlife species—these are chilling, sobering thoughts. Any ethical person will respond to them with a kind of horror. But there are warnings that are beyond right and wrong. One of the report's findings was that our ancient connection to our hunting culture may well disappear—within my grandson's lifetime. Knowing what price my grandchild was being asked to pay for other people's foot-dragging would give me inspiration for many years to come.

It might not have been as ambitious as many of us would have liked, but in light of what was happening to our homelands because of climate change, this document meant a great deal to me, in more ways than one. It would now become a key resource we could use to signal to the world how dire the Arctic situation was—and how we urgently needed to address climate change. It would be instrumental in building a strong political strategy for combating climate change. Now, thanks to the hard work of our hunters and elders, the scientists, the circumpolar Indigenous leaders and staff, and the politicians and their bureaucrats, the void in Arctic climate change science was filling up. The Arctic sentinel now had a powerful tool for engaging in an even stronger campaign to save our home.

7

The Right to Be Cold

THE WAR CRIMES AND ATROCITIES of the Second World War drew the Western world's attention to a profound gap in international law. While the concept of legal, political, and ethical codes of conduct and natural laws had existed for thousands of years, across many cultures, the idea that there were basic rights that all people in every nation and of every political or religious affiliation should share had never been internationally recognized. After the war, many felt that there needed to be an agreement that outlined and enshrined these rights in law.

The newly formed United Nations began working on such a document, and in 1948, adopted the Universal Declaration of Human Rights. Two additional covenants were added to this declaration in 1966: the International Covenant on Economic, Social and Cultural Rights, which identified labor rights and the right to health, to education, and to a sufficient standard of living, and the International Covenant on Civil and Political Rights, which included such things as freedom of religion, of speech, and of assembly, and the rights to vote and to have a fair trial. These three documents constitute the International Bill of Human Rights. While not a formal treaty, by 1976, enough countries had ratified the International Bill of Human Rights to allow it to come into force as international law.

The covenants reflect the fact that the human rights outlined in this work by the United Nations fall into two basic categories: civil and political rights, and economic, social, and cultural rights. Although many countries tend to emphasize one category of rights over the other, it's generally agreed that the categories are both indivisible and dependent on one another—you can only secure economic, social, and cultural rights if you have civil and political rights; likewise, you can't adequately exercise your civil and political rights if you have no economic security or social freedom.

Since the establishment of the Declaration of Human Rights, other human rights bodies have been created, including the Inter-American Commission on Human Rights (IACHR), based in Washington D.C., which was founded in 1959 and which grew from the American Declaration of the Rights and Duties of Man (a document that actually predates the UN Declaration).

Above all, then, a right is a protection against the power of others, whether or not that power is wielded maliciously. In the past, Inuit populations had seen their rights trampled. Now, with our world melting around us, we were again experiencing this assault against our rights. Our work with two U.S. environmental groups would help us make that case to the world, and would encourage the global community to recognize that environmental protection is intrinsically linked to the protection of human rights.

While working on the Arctic Climate Impact Assessment, those of us in the ICC and in Inuit communities had come to

know two things: first, something had to be done about the
issue of climate change, and second, this challenge, involving
every country of the world and global industry, would be
much greater than negotiating the POPs treaty. The ICC had
attended two of the United Nations' Conference of the Parties
(COPs) meetings since the Earth Summit in Rio, but I had not
been heavily involved in this process. The politics of the ACIA
process, however, had shown us the hurdles we would face in
the larger battle against climate change.

In the early 2000s, ICC Canada held a board meeting
in which the boards of directors who headed the regional
bodies—that is, the Makivik Corporation, Nunavat Tunngavik
Inc., Inuvialuit Regional Corporation, and the Labrador Inuit
Association (now the Nunatsiavut Government)—talked
about their concerns and the possible approaches we might
take to share our experience of climate change and ask for
environmental protection.

We all knew that while we in the Arctic were witnessing
the effects of environmental degradation every day, for many
around the world, climate change was still an abstraction.
This point had been driven home for me shortly after France
became the fiftieth nation to ratify the POPs treaty, making
it enforceable. I was holding a climate change workshop in
Iqaluit and invited John Buccini, the Canadian who had
chaired the intergovernmental POPs-negotiating session,
to thank him for bringing the Stockholm Convention to a
resolution. At our event, he attended a meeting I was chairing
to update regional leaders on the ICC's international efforts
to address climate change. At the end of the day he told
me that, until this session, he hadn't recognized the strong
parallel between the POPs issue and climate change for Inuit.
John's admission made me realize that we couldn't assume

that our global leaders or the population at large would see the connections between toxins and climate change, or that they would understand how many things were affected by our changing environment.

Indeed, the majority of the world's population now lived in cities, and people in urban areas were often far removed from the land that supported them. A great disconnect had grown between city dwellers and the environment. Too many didn't realize that the cars they drove and the emissions they created by powering their cities were connected to the Inuk hunter falling through the thinning ice, and to the Pacific islander defending his home along the sinking shore. For cities to reflect true ecological integrity, those who lived within them needed to look inward to realize the effects of their decisions on urban populations, but also outward to understand how their decisions affected the entire world. We believed that once city residents realized this profound interconnectedness, they would be able to relate to vulnerable communities around the world, as a shared humanity.

But how could we Inuit and the ICC get this important message across?

During one of our ICC Canada board meetings, Pita Aatami, president of Makivik, wondered what we needed to do to draw the world's attention to the imminent devastation of the Inuit way of life. He asked, "With our food being poisoned with toxins, our ozone layer being depleted and now climate change affecting our hunting culture, what strong stance do we need to take to be heard? Do we need to launch lawsuits to draw attention to this serious matter?" My mind began to spin as I thought about this proposal; not only would the legal process be extremely expensive but also the world might interpret this action as a pursuit of money. I was vehemently opposed to this becoming an issue of money. I've always felt that whatever action we take, whatever path we decide to follow, we must remain on the moral high ground. I thought we could make a stronger impact through the politics of influence than through the politics of conflict or confrontation. I'm a firm believer that synergy is created when you look for answers that will bring about a change in perspective at a time when people, the world, cultures, or communities are ready for that change.

Luckily for us, the Inuit and Arctic communities weren't the only ones grappling with how to get the world's attention focused on climate change. A number of environmental law institutes in the United States had already been thinking about legal strategies that might put pressure on governments of the world to address environmental issues. Their focus had been on the United States, in particular, as it had been almost completely absent from the UNFCCC process and the efforts to lower greenhouse gas emissions. It had become evident that the world's leading economic power and worst greenhouse gas emitter (back then) was reluctant to lift a finger to do anything about climate change.

One of the environmental institutes looking at legal avenues was Earthjustice, based in San Francisco. Attorney Martin Wagner from Earthjustice and Donald Goldberg from the Center for International Environmental Law (CIEL) in Washington, D.C., had been working together to link the issue of the environment to human rights. It was a brilliant approach.

Martin and Donald felt that the human rights discussion was due for an update. They had recognized that in an era of striking environmental damage, people's economic, social, and cultural freedoms were affected not only by their civil or political freedoms, but also by the changing climate and environmental degradation. To secure the already-recognized human rights, populations would also have to be protected from devastating environmental change. Another way of looking at it was that since it was proving difficult to protect the environment, perhaps they would have more success protecting the people in it.

Martin and Donald had heard about my work on the POPs treaty from Dan Magraw, who was heading CIEL at the time, and whom I had met during the negotiations. They invited me to meet with them, hoping that they might find support in the Inuit community for an effort to link Arctic climate change to human rights. Terry Fenge and I were in Washington, D.C., in late 2002 to attend a few events, and we met with Donald for the first time in a hotel lobby in the capital. I put a great deal of weight on first impressions, and I wanted to see if Donald's approach would be compatible with Inuit interests and concerns and whether he was the sort of person we could work with. While our meeting was brief, I left feeling comfortable that Don was knowledgeable, but just as importantly, well intentioned—a man with his heart in the right place.

A few weeks later, we had a conference call with Martin from Earthjustice to further explore this avenue. In our talks, Donald and Martin explained that under the structures of international law, a human rights petition is really the only way for a non-governmental group (or an individual) to directly address the world's governments, and that legal petitions are also a powerful way to convey the human story of an issue to a global public. They suggested that a human rights petition—a lengthy document that would include science, research, witness statements, and legal arguments—would allow Inuit to tell the world what was happening to them, to put what had been a largely scientific debate into human terms. Given how quickly and unequivocally Inuit culture, health, and economic well-being were being affected by climate change, we were ideally equipped to link climate change to basic human rights—to argue that the protection from climate change was essential in order to secure the social, cultural, and economic rights that were already internationally recognized.

Donald and Martin were planning to focus their petition on the United States. Not only was it the world's largest producer of greenhouse gases but, just as important, it also had not been supporting the international treaties on environmental protection. While other countries, like China, were also producing CO_2s in serious amounts, Donald and Martin felt that the developed countries should be first in line to bring their emissions down. And there was also a powerful and progressive human rights tribunal that had jurisdiction over the United States: the Inter-American Commission on Human Rights. They explained that the European Court of Human Rights was also good, but it did not have jurisdiction over the United States. The IACHR had also, apparently, already started to make links between the environment and human rights in some of their decisions, so it looked as if they might be the most open to hearing a petition like ours. (I would later learn that the United Nations Commission on Human Rights [UNCHR] was embroiled in controversy at the time, with a number of member countries being accused of significant human rights violations. It was probably not, therefore, a viable option for our petition. The UNCHR was actually disbanded in 2006 and replaced by the United Nations Human Rights Council.)

After our meetings, Terry and I went back to our respective homes and our other work, mulling through everything Martin and Donald had proposed. In Iqaluit, I continued to educate myself on the petition process, and spent many hours trying to assess if this was something that I should propose to the ICC Canada board of directors and the executive council of ICC International. Given that I would probably be leading this campaign, I also spent some time thinking about whether I was willing to champion the cause through the petition process. One thing I had learned as I came to fully understand the implications of launching a human rights petition: this was going to be a much bigger job than anything I had done before. But despite that, I made the decision that this was the right thing to do.

In the meantime, Terry Fenge, in Ottawa, had been weighing all this himself, and talking in further detail with Martin Wagner and Donald Goldberg. He felt strongly about going forward as well. In February 2003, while I was in the Arctic, I received a long memo from Terry. In it, he recommended that ICC Canada and ICC Alaska (the two countries that could use the IACHR) submit a petition to the IACHR seeking an authoritative declaration from the commission that the impacts and effects of human-induced climate change in the Arctic amount to a violation of the human rights of Inuit. He noted that the ACIA report could provide the crucial data base and analysis to establish what was happening in the circumpolar North.

After much thought, and with a deep sense of responsibility, I prepared to present the idea of a human rights petition to the upcoming executive council meeting in Ottawa. Unfortunately, I hadn't noticed the cc list in Terry's memo, thinking it was addressed only to me. In fact, it had been copied to Stephanie Meakin and Corinne Gray, the executive director of ICC Canada. What happened next took me by surprise.

In late 2002, Martin Wagner and Donald Goldberg joined me in Ottawa at the ICC International executive council meeting to present the human rights petition idea and to ask the council for their support. In particular, we were looking for interest from the Canadian and Alaskan members, as Greenland/Denmark and Russia were not signatories to the Organization of American States, under which the IACHR operates. The reaction of the Canadian executive council however, was unexpected and alarming. They were skeptical and suspicious of the process. One member muttered under his breath, "This is not good." They were not open to even exploring the proposal. In the end, I had more support from ICC Greenland and ICC Alaska. It was a disappointing turn of events.

When I finally figured out that the Canadian members had read Terry's memo before the meeting, I began to understand that distrust. They associated the idea with Terry, an outsider. There is often a challenging divide between Inuit and non-Inuit "experts," "advisors," environmental groups, or conservation NGOs that has arisen from a long history of non-Inuit coming into our world and creating policies that eventually hurt our communities. All too often, those who are out to save the world are all too ready to sacrifice Inuit and our way of life. Greenpeace and other animal rights groups' campaigns to stop the seal and whale hunts have devastated the livelihoods of Arctic Indigenous peoples, and had ignored our culture's respect for these animals. Resentment toward these conservation NGOs has spilled over to other *qallunaat* organizations. Even before Terry sent the memo, some members of ICC Canada were always going to be suspicious.

But there were other factors in the resistance the ICC presented to the human rights petition. Some people were afraid of repercussions from governments, including from the United States. One of our Canadian national members thought the Canadian government might cut our funding if we moved ahead with this endeavour. Yet we had, and would have for many years to come, a transparent, productive relationship with the government in Ottawa, notably with Minister of the Environment Stéphane Dion and Foreign Affairs Minister Pierre Pettigrew. Also, I shouldn't present the challenges as a fault line between Inuit and the rest of Canada. There are plenty of rifts *within* the Inuit community as well. (As we moved forward in the process, I would again face resistance in my own backyard. Two Inuit officials within the federal government resisted my work at different times. One went so far as to say, "I don't want to be seen scheming with Inuit." The other tried to convince my ICC board of directors [when I wasn't present] that I was overstating the issue of climate change and that we just needed to focus on adapting to the changes.)

While I initially felt rather alone in the campaign, I pressed forward and eventually got enough written support in the form of resolutions from the board members within Canada and the executive council members of ICC to continue to develop the petition.

After the ICC International meeting, Terry, Martin, Donald, and I looked at how the campaign would fit in with the UNFCCC process and started to plan for the next Conference of the Parties meeting, COP 9, in Milan, Italy, in December 2003. These COP meetings have been held annually since the UNFCCC was entered into force to assess progress on climate change and make new commitments. Martin and Donald started writing the petition itself. They drafted the legal argument with help from environmental lawyer Paul Crowley, who had been working with me since partway through my term as ICC International chair. Residing in Iqaluit for a number of years with his family, Paul served as my legal counsel and adviser, demystifying some of the legal jargon and process for me as I took on the role of political head of the petition process. (Paul also ended up writing the executive summary for the petition.)

Our plan for Milan was to announce to the world that we would be launching a human rights petition on climate change that would target the United States and their greenhouse gas emissions. We wanted to gauge how the world would react to this route. We had arranged to have a side event (a two-hour presentation outside of the plenary session) at the meeting where Don, Paul, and I would present this human rights strategy, but we also hoped to get the Arctic and its environmental concerns raised in the full-member (or plenary) sessions as well.

Milan was the first UNFCCC COP that I had ever attended, and I wasn't quite prepared for the size or the feel of the conference. Rather than hundreds of attendees, as in the meetings on persistent organic pollutants, COPs attracted thousands of people. And yet there was no perceivable heartbeat to many of the issues or meetings. Instead, there were scores of bureaucrats working away furiously on their interventions (presentations) and strategies, writing and editing without, apparently, much thought to the fact that these documents had anything to do with people and their lives. Our own Canadian delegation, sadly, were initially no exception.

While we had established a strong connection with some of the top government officials handling the POPs treaty, we didn't have this relationship with the Canadian delegation for the UNFCCC process. When I first arrived at the Milan COP, I approached the head of the Canadian delegation and asked if I could be allowed to make a brief intervention on the plenary floor. She quickly replied that interventions could be made only by heads of state. I then asked if she could make a couple of statements on behalf of the peoples of the Arctic, or if she could inject the word *Arctic* into the Canadian interventions. I reminded her that those who call the Arctic home are negatively and disproportionately affected by climate change. She told me that the Canadian delegation was participating in only two workshops at this COP, and they would not be mentioning the Arctic.

Rebuffed by our own Canadian government officials, we began to look for other ways to get mention of the Arctic into the plenary presentations at the COP. Paul Crowley had been working the halls and lobbying with various other countries when he connected with the delegates from Samoa, another country that was being affected by Arctic climate change. Rising sea levels created by the melting of the polar ice were threatening their shorelines and coastal communities. The Samoan delegates agreed to make clear the Arctic connection in one of their interventions.

There you have it. We had been forced to depend on a tiny country, far away from the Arctic, to remind the world of our existence when our own country would not.

But the tide turned when we held our side event. It was a crystal-clear evening that night in Milan, and a full moon hung over the city. As Donald, Paul, and I headed into the conference room we had booked for our presentation, we were shocked to see that it was full. Not just full—overflowing. On this beautiful winter evening, our stuffy meeting room was standing room only, and people were spilling into the hallway. We could see them outside the doors, peering in, trying to hear what we were saying.

We walked the audience through the ACIA science and the dire environmental conclusions that had been raised in that report. But we painted the human picture as well. We talked about satellite measurements of retreating sea ice, but we also told stories about how precarious the sea ice had become for the hunters and how, in many areas, the melting tundra was causing our homes to crumble and collapse. We described the changing reality of Inuit life and the human suffering that accompanied the melting Arctic.

The audience responded enthusiastically. The power of the rights-based approach was that it moved the discussion out of the realm of dry economic and technical debate that too often overtakes discussion at UN climate change conferences. Instead, this new strategy took the path of principle, showing that fundamental change was not just sound policy, but an ethical imperative. That approach seemed to resonate powerfully with the UN participants.

We had cast our line to see what fish we would catch, and instead we caught a whale. Delegates and other attendees stopped us in the hallways and meeting rooms to talk more about our plans. Journalists from around the world approached us for interviews. Our story was covered by news outlets from across the United States, Europe, Asia, the Middle East, and, of course, Canada.

It was during one of those interviews that we found the phrase that would capture the heart of our struggle. While still at the Milan convention, I was being interviewed by a journalist from one of the major U.K. newspapers. I was explaining how Inuit culture and economic independence, as well as Arctic wildlife, depended on the cold, the ice, and frozen ground, how the great shifts in temperature and weather patterns were upending an entire way of life, and denying us our economic, social, cultural, and health rights. "Ah," he said. "You are fighting for the right to be cold."

That's it, I thought. *The right to be cold.* From there on out, it became the phrase we used to capture our struggle.

("The right to be cold" was a wonderfully evocative expression, but it was open to misunderstanding too. Years later, at a meeting of Canadian judges hosted in Iqaluit, one of my colleagues told me that she had just been talking to a woman about me. "Oh, is she the one who is fighting for the right to be cold?" the woman asked. "Well, what about *my* right to be warm?" She may have been joking, but I so wished that she had said this directly to me. If she had, I would have noted that she was already warm. And, living in the South, was likely to get a good deal warmer still in the coming years. But I would have also explained that the right to be cold doesn't refer to individuals. It refers to the circumpolar environment, the Arctic land and the way of life that depends on ice and snow. And while she may really have not understood what we were saying, I've also noticed that some people seem to deliberately misunderstand as a way of dismissing our demands. The rights we're fighting for are *her* rights too. Just as our environment is her environment too. That's why we want people like her to join us. We all have the right to be protected from climate change.)

Building on the momentum of the COP meeting in Milan, Paul Crowley and I attended the next COP meeting a year later in Buenos Aires, Argentina, along with Sasha Earnheart-Gold. Sasha was a student from Dartmouth College in New Hampshire who had been researching human rights in relation to environmental degradation when he discovered our campaign. He offered to work for us for free (with the help of a college fellowship and funds he'd help raise). Being a small NGO with limited resources and staff, we were pleased to have him.

The media interest had been building since the Milan COP. By the time we got to Buenos Aires in December 2004, we knew we were on the world's radar, particularly that of the United States. What we didn't realize, however, was that not all of the attention we were attracting would be positive.

We had planned to have another side event in Buenos Aires, which we were promoting with posters that we'd placed around the COP venue. On the day of our event, we realized that someone had gone around with a red marker and defaced many of our posters. That night, as we presented an update on how far we'd gotten in preparing for our petition, we noticed that two men seemed to be placed strategically in the audience. I recognized one from other events. They repeatedly stood up to question our science, often saying that the facts we were presenting were unproven. The two also argued with the route we were taking and questioned our motives. It was too coordinated to be happenstance. It seemed clear to us that most of those challenging us were from right-wing organizations from the United States. It was a trying media event.

But the upside was that if our work was generating defensive attitudes in the United States, we had made our mark. And in spite of the attacks, the Buenos Aires COP did allow us to find further support. I had a productive meeting with the Canadian minister of the environment, Stéphane Dion, who was extremely open to hearing about our strategy with the petition. He and his lawyers were impressed with the direction we'd chosen, calling it "brilliant." I also spoke with the leader of the federal New Democratic Party, the late Jack Layton, who was always an active supporter of our work and cared deeply for the environment. (In later years Jack would ask me to run for federal politics. I graciously declined, knowing I couldn't possibly survive the loud and uncivil manner in which the House of Commons conducts itself.)

Buenos Aires also gave us an opportunity to talk further about our mission. I spoke on three occasions at this COP session, and each time I would refer to the objective of the UNFCCC that had struck me so strongly at the beginning of our ACIA work: "the stabilization of greenhouse gas emissions ... that would prevent *dangerous* anthropogenic interference with the climate system" (italics mine). I would remind the audiences that we as Inuit were already at that threshold of *dangerous* interference. We were not preparing for it, we were *living through it*, and it was getting worse.

As the week unfolded, we became aware that it wasn't only the right-wing American interests who had concerns about our work. Some delegates at the conference appeared to feel that we were pushing too hard and were too narrowly focused on a single group of people. In other words, we were accused of being selfish. Our opponents seemed to be implying that we were trying to save the world exclusively for ourselves. But you can't save the Arctic on its own, any more than you can spew POPs or CO_2 in your own backyard and expect it to stay there. It's *all* connected. I responded by saying that we were not working solely for Inuit interests. I reminded people that the planet and its people were one, and that our struggle was inclusive by nature. "By protecting the Arctic, you save the planet" became one of my mantras. But I did admit that we were pushing hard. We had to. My biggest worry and the driving force behind everything I did was that I didn't want my grandson to look back and ask, "Did she try hard enough to get the world to understand and take action when she was in that position of influence?"

But the pushback we were getting wasn't coming just from the outside. Our own Inuit camp was also engaged in ongoing politicking. This became evident as we prepared for

COP 11, which would be held in Montreal from November 28 to December 9, 2005. Unlike the COP events in foreign countries, which only I or an elected ICC official, along with our staff or advisors, could attend, COP 11, being in Canada, allowed for many more of our board members and leaders from our Inuit communities to join us. But as we worked through who would be attending and how our side event would run, it became obvious that support from the ICC Canada board of directors and certain executive council members was not a given. Some of our regions were moving in the direction of resource extraction or gas and oil pipeline businesses. (I talk more about our communities' embrace of these industries and the challenges that presents in Chapter 9.) ICC representatives from those areas were fearful about the implications of being signatories to the petition. They did not want to look like hypocrites.

I tried to convey to them that we weren't asking to stop all development, but rather to move in a direction that would be more sustainable—not just for ourselves in the Arctic but also for the whole world. While they continued to approve the work I was doing as ICC chair on the petition, it appeared difficult for some of them to embrace the idea that the petition would not threaten development in the North.

My struggle to get our leaders to feel safe supporting the petition tested my resolve over the weeks and months, and took a toll on me. Others on our team noticed. At one point in 2004, I was in the southern United States making a pitch for financial support for ICC's work at an international funders' meeting, when a call came through from Terry and Paul. They explained that they felt I was being isolated among my colleagues and that it was going to be too difficult for me to carry on alone, considering the challenges I was facing from my fellow leaders. They thought that perhaps I should abandon the petition approach.

I appreciated their concern, but I knew I would not be giving up this work on human rights and climate change. I had just received some terrible news about a young person I knew well. This fellow Inuk had been the victim of an act of horrific violence. The news was, of course, a shock. How could it not be? A young person so senselessly and so savagely assaulted. But it was also a heart-rending echo of so much of the news coming out of our communities—suicides, overdoses, abuse of women and children, alcohol-related accidents. Violence had come to plague our communities. Yet we had a right to safety and health, just as we had a right to a healthy environment where we could continue in our wise and patient hunting culture, a culture that demanded morality, a strong sense of self, and a life without such violence.

I was in tears as I told Terry and Paul about what had happened to this friend. I also told them that the difficulties I was experiencing with our campaign were nothing compared to what this young person had just survived. What I didn't say, however, was that for days, I had found myself imagining the terror of the assault, and amazingly, it gave me strength. I *knew* that the work we were doing might be the best way to protect my community from more tragedies like this one. In fact, hearing the news of that young person's ordeal was the moment I decided that I would never give up, that I would continue to lead with love and not fear. Love for children affected, love for my culture, love for our planet. Terry and Paul knew then that they didn't need to worry about me.

Meanwhile, Paul, Donald, and Martin continued to finalize the petition. Using the scientific conclusions detailed in the ACIA report, the petition presented extensive evidence about

climate change and warming. It also featured several pages
that documented the United States' role as the world's largest
contributor to CO_2 emissions. The document discussed how
existing American environmental laws were ineffective in
reducing greenhouse gas emissions, and how current U.S. law
could not provide sufficient or effective protection against the
human rights violations that Inuit were experiencing.

At the core of the petition were the numerous ways in which
Inuit human rights were being violated by climate change. The
petition pointed out that the American Declaration of the
Rights and Duties of Man (the document that serves as the
basis for the IACHR), as well as other international instruments,
protects rights "to the benefits of culture, to property, to the
preservation of health, life, physical integrity, security, and a
means of subsistence, and to residence, movement and the
inviolability of the home." Based on this, the petition pointed
out that "because Inuit culture is inseparable from the condition
of their physical surroundings, the widespread environmental
upheaval resulting from climate change violates Inuit's right to
practice and enjoy their culture." The petition also argued that
we were being denied the right to use and enjoy our traditional
lands, as the land was either changing or becoming inaccessible.
The fact that we were unable to hunt as before for food and for
hides and skins for clothing and that the loss of ice and snow was
damaging our snow machines, our sleds, and our other tools was
a violation of our right to personal property. The Western store-
bought diet we were being forced to adopt, the accidents caused
by melting ice and snow, and our increasing exposure to UV
radiation, among other things, meant that our rights to health
and life were being severely constrained. The petition also stated
that our fundamental right to residence and movement was
being violated as our homes were damaged and the land upon

which many of our communities were built was being eroded by melting permafrost. And finally, Inuit's fundamental right to their own means of subsistence was being denied as climate change was hurting almost every aspect of our hunting culture: the quantity and quality of wildlife, the length of the hunting season, methods of traveling, and the ability of our elders to pass on traditional knowledge.

To prepare for the arrival of the two young American students who would be gathering the testimonies, I had contacted all the selected communities they would be flying into to inform the leaders either through the community radio or by calls to the mayors or presidents of the regions and/or wildlife organizations that I needed their guidance on who should be interviewed. I explained what the petition was about and how important the voices of the communities were to this process. So the lifeblood of the document became the Inuit voices that told our story in heartbreaking detail. Sasha Earnheart-Gold, a Dartmouth student, and Rich Powell, a Harvard student, had traveled into the northern Inuit communities in Canada and Alaska to gather and film testimonies from Inuit elders and hunters. Before Rich and Sasha headed out with their video cameras and copies of the mock petition, lawyer Sandra Inutiq (originally from Clyde River, and a dedicated advocate for the protection of the land and Inuit culture), Martin, and I met in my Iqaluit home with the young men. Sandra and I discussed with Rich and Sasha the need to explain to the elders and hunters what the petition was all about and how their testimony would be used in it. We instructed them to make clear to the elders the importance of their knowledge—that it would be their expertise, their observations, that would inform our document. We also had Rich and Sasha ask each person interviewed if he or she would be willing to join me as signatories on the petition.

Rich and Sasha's work contributed enormously to the power of our petition. They brought the eyewitness accounts of sixty-three hunters, elders, and women to our cause. (I quoted a number of these wise people in Chapter 6.) What's more, after the filming was completed, Rich and Sasha told me that only one of the elders had declined to sign the petition. When I heard that sixty-two fellow Inuit would be joining their names with mine on the petition, I was overwhelmed. I sat on my living-room couch, looking out at my Arctic view, and the tears came. A sense of relief, comfort, and reassurance washed over me. I was not alone. The co-operation of Inuit from both the United States and Canada, as well as the involvement of Sasha, Rich, Donald, Martin, Paul Crowley, CIEL, and Earthjustice, underscored the message of connectivity and common humanity that we hoped to share with the wider world.

The COP meeting in Montreal was a two-week-long event. We would be launching the petition in the second week. Leading up to the launch, however, were numerous UN and media events focused on the Arctic and Inuit organizations and communities. Stéphane Dion, who impressively chaired the COP, spearheaded an Arctic Day event. Many of our own community members, elders, youth, and performers were there, promoting the Arctic way of life in our country. A great number of media events were held, at which we showcased the ICC, the ITK, and our own regional organizations. To be doing that in our own country, not so far from our own Inuit homelands in the North, made the COP a remarkable experience for me.

While many Inuit organizations got media attention during the event, as the political head of the petition group, I received a steady stream of requests for interviews and appearances, all deftly managed by Rich Powell, now my executive assistant.

Documentary filmmakers from France and Canada followed me from the time I landed in Montreal through the entire two weeks, recording my participation in various modules, speaking events, and interviews. Wherever I went I was surrounded by cameras and sound equipment and people. By the second week, when anyone asked, "Has anybody seen Sheila?" the reply would be, "Just follow the filmmakers, and you'll find her."

Before COP 11 had begun, Terry, Don, Martin, and I had worked to create a panel of respected authorities to accompany me at the official launch event. Foreign Affairs Minister Lloyd Axworthy and our own Mary Simon had been strong contributors in the creation of the Arctic Council. As ICC Canada president, I'd been working with Lloyd within the Canadian government from a distance, and I knew that he had maintained a high regard and respect for the Arctic and its peoples.

I asked Lloyd if he would be the moderator for the panel at the launch. After I told him about our petition work linking climate change to human rights, he shared the story of his own work on land mines. As the minister of foreign affairs, he had spearheaded a global treaty banning anti-personnel mines (known as the Ottawa Treaty). He explained that they hadn't made much headway on the issue of land mines until they incorporated the human rights angle into the debate. He understood the strategy behind making climate change a human rights issue and was an ideal person to moderate the panel.

The panel itself consisted of Martin Wagner, Don Goldberg, fellow petitioner Jamesie Mike (an elder from Pannituuq), and James Anaya. James was a renowned Indigenous human rights lawyer with the University of Arizona and is now the UN Special Rapporteur on the rights of Indigenous peoples. James is a remarkable man and, as a fellow Indigenous person, an important contributor to the discussion. I felt a great deal of comfort having him and Jamesie close to me as we launched the petition.

This support turned out to be crucial for me, as the day before the official launch, a launch we had been planning for months, our team received some bad news. In the weeks before the Montreal meetings, the ICC had started to waver, saying that they weren't sure if they could sign the petition. The day before our presentation, the team was in my hotel room in Montreal, working on the last resolution we needed from the ICC, when we had a conference call with the Canadian board of directors. At this final hour, the ICC backed out, saying that they would not be signatories. However, led by highly seasoned politician Nellie Cournoyea, a former premier of Northwest Territories and now the elected chair of the Inuvialuit Regional Corporation, the ICC Canada board of directors passed a resolution stating that the sixty-two hunters and elders who had chosen to become signatories alongside me would have the ICC's political support as we moved forward.

After all the time I'd spent hoping for ICC Canada to join me in this historic moment, I would have expected to feel disappointed by the turn of events. Instead, I felt relief. I would no longer be wearing institutional shackles. I would be free of internal ICC politics. And I *did* have the support of sixty-two fellow Inuit. I would be able to move forward in this human rights work in the way I felt I needed to. I always say that all things have their way of unfolding. This was as it was meant to be.

Despite the sense of freedom I now felt, the morning had been difficult, and the challenges were compounded by the relentless media attention. Journalists and TV crews had been with us throughout the entire process. But they had been covering the story as a David-and-Goliath tale, which I found frustrating. I had never felt that this was a fight. We were reaching out, not striking out. We were working from a position of focus and strength, not victimhood. We wanted to educate and encourage the U.S. government to join the global effort to combat climate change. In a very real sense, our petition was a gift from Inuit hunters and elders to the world and to those most negatively affected by climate change. It was an act of generosity from an ancient culture deeply tied to the natural environment and still in tune with its cycles and rhythms. It was a gift to an urban, industrial, and "modern" culture that had largely lost its sense of place and position in the natural world. I just hoped that after our official presentation, this point might become clearer.

I had told my team that I didn't want to sit in the conference room watching as it slowly filled up with people. Instead, I planned on entering just before the event was to start. Just minutes before the hour, the two documentary cameramen, Rich, and I walked the long corridor that joined the hotel to the Palais des congrès de Montréal, where the COP was taking place. I met Lloyd Axworthy along the way. Despite the presence of Rich, Lloyd, and the cameramen, I was aware that in many ways I was alone. I remember thinking, *I have to get this right. I have to look strong, and I have to feel strong.*

The room was packed once again. Media outlets from around the world were there: BBC World, Reuters, the *Toronto Star*, Al Jazeera. Many people from the Arctic communities were there as well, including Jamesie Mike and his daughter, Meeka Mike. My brother Charlie, a long-standing Liberal senator, was sitting near the front. (I mentioned during the presentation how he had led our dog teams when I was a child. After the event, he told me how much this had moved him.)

His presence in that room was a huge comfort to me. It would have been easy to have been overcome by the importance of this launch and what was riding on our petition. Having family in the room as I was about to describe our efforts to defend our rights as an ice-dependent people made me feel stronger.

Then our presentation and panel discussion began. While the vast majority of the audience seemed to be thoroughly engaged and responsive, about halfway through the discussion, I noticed some of the media ducking out, including journalists from Reuters and the *Toronto Star*. They would later say that they had headed over to other conference rooms to gauge the reaction of the American representative to our petition. When the launch concluded, we were again surrounded by journalists and cameramen, asking for interviews, but those who had left early never came back. I would later realize that this was a sign of things to come.

The Montreal COP ended on December 9. Before it concluded, on December 7, 2005, Paul Crowley, on my behalf, submitted our petition, "Seeking Relief from Violations Resulting from Global Warming Caused by Acts and Omissions of the United States" to the Inter-American Commission on Human Rights.

And then we waited.

IT WAS A SURREAL TIME. After the crush of interest leading up to and during COP 11, the media all but disappeared. In early February Donald sent the team an email reflecting on our work up to this point and thanking everyone for their efforts. He noted that our announcement of the petition at COP 9 seemed to "dispel fear and renew hope" in people. The message, he said, had not abated. Only the day before he'd received an email from a student who had just learned about the petition and was "awestruck."

He also had kind words for me: he called my efforts Herculean and said, "Few people have the stamina, dedication, perseverance, concern and love that was needed to take us as far as we have come. So many times, I thought to myself, the odds against us are insurmountable. Sheila will never overcome the problems, impediments and forces arrayed against her. Clearly, I didn't know what you are made of."

Donald's words were important to me, not because they stroked my ego, but rather because they commented on how hard I had been working and how tough everything had been. I found it oddly affirming that someone else on our team had had doubts about whether we would get to this point. His letter acknowledged that the struggle had really been as tough as it had sometimes felt to me, and it touched my spirit.

Donald also talked about the next phase of the campaign, referring to the delivery of the petition as a conclusion but not the end. But it *did* seem to be the end of the world's attention for a while.

Of course, none of us were sitting still while we waited to hear from the IACHR. While working on the human rights campaign, I had also been involved in a number of other initiatives. One of these was the United Nations' Many Strong Voices program, which was launched at COP 11 in Montreal as well. The program was based in Ottawa and run by Ilan Kelman of the Oslo-based Center for International Climate and Environmental Research, and John Crump, who had previously been the executive secretary of the Indigenous Peoples Secretariat in Copenhagen. Many Strong Voices linked the Arctic with the Small Island Developing States (SIDS), the designation for fifty-two island countries in the Caribbean, the Pacific, off the coast of Africa, the Indian Ocean, the Mediterranean, and the South China Sea. The program, established with the support of Klaus Töpfer, the executive director of the United Nations Environment Program, and his assistant, Svein Tveitdal from Norway (both of whom had been so helpful with our work on POPs), was developed to give Indigenous peoples adversely affected by climate change a voice, and to underline the connection between the melting Arctic and Greenland ice sheets and glaciers and the rising sea levels that were threatening so many low-lying coastal communities around the world. (The program continues to be run by GRID-Arendal and the University College London in conjunction with the UNEP. A further important initiative to connect Indigenous peoples affected by climate change was the Indigenous Peoples' Global Summit on Climate Change, held in Anchorage, Alaska, in 2009 and led by ICC Chair Patricia Cochran.)

Another bid to bring the world's attention to the melting Arctic had been the ICC's Arctic Wisdom event, a two-day briefing on scientific, cultural, and political issues related to climate change in the North, held in Iqaluit in April 2005. My friend David Veniot, who worked for Inuit marketing company Ayaya, and artist John Quigley had come up with the idea of having Quigley create an image in the Arctic that would send a signal to the world about the dangers of climate change. Los Angeles native John Quigley is an aerial artist who assembles people on the ground, usually on beaches in southern parts of the world, to create large images that are photographed from a helicopter. Most of his images carry a political, social, or environmental message. Over lunch in Iqaluit, David asked if I would be interested in leading this effort and hosting various players from Los Angeles, including some Hollywood celebrities.

The idea struck a chord with me. There were many months of preparation to secure funding and organize the event, which required community input and involvement from the city of Iqaluit and the schools, and in the end received financial support from a host of environmental groups and private businesses, many Inuit owned and operated. John Quigley and his team from Spectral Q, along with actors Jake Gyllenhaal and Salma Hayek, arrived in Iqaluit for the shoot on April 22. Others, including Tom Kelly of KyotoUSA and Matt Petersen of Global Green USA, also came up. (Tom Kelly would become an ally in the effort to get the United States to address their emissions. After his trip to the Arctic to partake in this event, we would often find ourselves at the same meetings and events.)

For the photo shoot we had a thousand Inuit, dressed in our northern clothing and fur, lie down on the ice of Frobisher Bay, off Baffin Island, to form an image of a drum dancer with the message: "Arctic Warning." The event garnered interest from American media outlets, including *People* magazine.

While we were waiting to hear from the IACHR, however, I had to respond to less-positive media coverage. On March 2, 2006, Paul McCartney and his then wife, Heather Mills, were filmed on the ice in the Gulf of St. Lawrence, petting a baby seal (Heather almost getting bitten in the process) to publicize their anti-sealing campaign. The next day they appeared on CNN's *Larry King Live* to discuss their anti-sealing work and to debate with Newfoundland premier Danny Williams, who tried to explain the importance of the hunt and how it has been misunderstood and mischaracterized.

Many of us in the Inuit community were deeply upset by McCartney's actions and words. The Beatles' music had played such an important part in our early lives. It had seen us through our school days away from home and through the changes that had come to our communities from the South in the sixties and seventies. We felt that Paul McCartney had let us down. Jose Kusugak, the ITK president at the time and the former drummer for our own Harpoons band in Churchill, wrote an editorial in the *National Post* titled "Inuit to Paul: Let It Be," in which he explained the Inuit perspective on the seal hunt and the damage done to our communities by misguided animal rights activists. Zebedee Nungak followed with a remarkable piece in *Inuktitut* magazine (issue 10, December 2011) in which he wove Beatles song titles into his narrative.

And, on behalf of all Inuit, the president of ICC Canada, Duane Smith, and I, the chair of ICC International, addressed McCartney and Mills's stunt in a press release. We wrote, "Speaking recently on the Larry King show, globally known singer and former Beatles member Sir Paul McCartney urged Canada to ban commercial seal harvesting, calling sealing 'barbaric' and 'archaic.' He suggested to Danny Williams, who was then the premier of Newfoundland and Labrador, that Canada should buy out the sealers and promote seal watching instead."

We continued, saying that lying down on the sea ice and playing with seals is, frankly, silly, and it's also disrespectful to wildlife. Seals may look cute, but they are not pets—they are animals that live in the wild. Inuit hunt seals for food and clothing, and we market internationally the by-products of our sustainable hunt. This is why attacking the commercial harvest on Canada's East Coast and attempting to destroy the market for seal products also affects the Inuit seal hunt in the Arctic. I added that global wildlife organizations, such as the International Union for Conservation of Nature (IUCN), note that seals are not endangered. The World Trade Organization allows unrestricted trade in seal products because seals are not listed as endangered under the Convention on

International Trade in Endangered Species of Wild Fauna and Flora (CITES). I also suggested that if Sir Paul really wanted to save the seals, he should help us stop climate change, which is destroying their habitat, the sea ice. Finally, we endorsed and repeated the invitation by Aqqaluk Lynge, president of ICC Greenland, to Paul McCartney to come to the Arctic to learn what seal hunting means to Inuit.

He never took us up on our offer.

The McCartney media blitz was a good reminder to me that the Arctic is better known for its wildlife than its people. Whether it's baby seals that need to be saved or corporations that market their products by capitalizing on wildlife, the world loves to focus on Arctic species that tug at the heartstrings of "Southerners." We see images of polar bears drinking Coca-Cola with seals, a highly unlikely scenario, as one is actually lunch, not a playmate, for the other. Many people in the South are sickened by pictures of animal blood on the ice, but to Inuit, this blood represents a powerful cycle of life giving life—an affirmation of life, not a confirmation of death. We Inuit of the Arctic have a profound understanding that this blood offered to us by our wildlife will keep our own blood warm and fuel us from the inside as we, along with our wildlife, spend hours in the deep cold—a deep cold that all life, including the flora, fauna, and Inuit, depend on in order to remain healthy and vibrant. As my friend Meeka Kilabuk once said to filmmakers who were in Iqaluit filming about the POPs issue, "Lipton Cup-a-Soup isn't going to keep you warm in minus forty, but seal meat will."

There is a story I often tell about Inuit affinity for the wildlife that sustains us. I once hosted a filmmaker from California—one of many filmmakers I've hosted over the years—while he shot a documentary about toxins in our marine life. He and his wife arrived at a time when my good friend Leena and I were eating some beluga *muttaq* for lunch. I offered him a piece, but seeing how his body leaned away from the bowl, I knew it wouldn't be an easy sell. While I'm not one to push issues on people, I wanted him to better understand our culture, so I arranged for him to go seal hunting with Ben Kovic. Ben was head of the Nunavut Wildlife Management Board at the time and was always willing to help me in educating those I would send his way for interviews or hunting trips. (I remain grateful to Ben, another former Churchill student, along with Iqaluit residents Pitseolak Alainga and Meeka Mike, for the time they took with Southerners so that they could gain a better appreciation of our culture.)

After a day of hunting on the water, Ben sent me an email saying, "Sheila, I think these people might be some kind of animal rights sympathizers, as they had a hard time looking as I shot the seal." I responded that this was a common reaction in people from the South. Later in the week, I told the filmmaker that I would have him eating *muttaq* before he left. He quickly responded, "Oh no, I can't do that. I have too much of an affinity for dolphins and whales to eat them." Without a second thought, I responded, "Ah, but we too have an affinity for whales, which is why we eat them."

The filmmaker paused, and in that pause was a palpable moment of reflection and insight—a true "aha" moment—that I had been hoping he would get before his departure.

A day or so before the filmmaker's flight home, we were filming a young mother, Lucy Qavavaug, who was nursing her newborn. Beside her on the coffee table were small slices of *muttaq*. Just as he finished filming, Lucy offered him some. Before he could think his way out of it, he spontaneously popped a piece in his mouth and started to chew. I immediately felt the energy of the room shift. On some metaphysical level,

I could almost hear the whale say, "Yes!"

What a moment! My only regret was that this powerful connection wasn't captured on camera.

But this kind of public education can be an uphill battle. McCartney's campaign had generated significant attention and garnered a great deal of sympathy from emotionally driven animal rights activists. In early April, just as I was leaving to attend the Global Green USA Green Cross Millennium Awards, where I was to receive the International Environmental Leadership Award, the Sea Shepherd Conservation Society launched a campaign on their website to try to get people to boycott the event. They had heard about the press release that Duane and I had sent in response to Paul McCartney's anti-sealing efforts, and clearly felt that I shouldn't receive the award.

Matt Petersen, president of Global Green USA, called me to discuss the kerfuffle that was starting in Los Angeles because of the boycott (which was later abandoned). I assured him that our press release hadn't been antagonistic, but rather had been an effort to educate people about sealing. In spite of the boycott attempt, the awards event turned out to be a wonderful opportunity to further educate the audience and the wider Hollywood community. While there, I was asked to be interviewed for *The 11th Hour*, a documentary about the state of the global environment produced by Leonardo DiCaprio and Leila Conners Petersen of the Tree Media Group, which I happily agreed to.

IN THE MIDST OF WAITING to hear back from IACHR, I came to an important personal decision. I decided not to run again for an elected position with the ICC, effectively leaving the organization in the summer of 2006. I was worn out from eleven years of almost constant travel and physically and emotionally taxing work for the ICC. But I was now without an institutional base, with no staff and no funds to carry on any substantive work for the petition, or to continue the human rights campaign. While the choice to leave had been mine, once I finished my term I felt frustration and a sense of paralysis.

Now living in this sort of limbo, it was hard not to notice month after month passing without word from the IACHR. Then, in early December, almost a year after we had submitted the petition, one of the lawyers from CIEL in Washington was told by a staff member from the U.S. State Department that the IACHR would not hear our petition. What a way to find out!

Shortly after this, an official letter came from the IACHR stating that it could not process the petition at present. The only explanation given was that the petition didn't enable the commission to determine whether the alleged facts would characterize a violation of rights protected by the American Declaration. The commission's letter did not question the essential facts: that global warming is threatening the lives, culture, and property of Inuit in truly devastating ways. In fact, in the year since the petition was filed, accounts from our people and new scientific studies had confirmed that the Arctic was warming at an unprecedented rate, and that this warming was attributable to anthropogenic emissions of greenhouse gases.

Upon hearing the decision from the IACHR, I immediately wrote to them, saying that we would provide whatever other information they needed in order to make clear the connection between climate change and human rights violations in the Arctic. I also sought a detailed explanation of their concerns. Without such an explanation, we wouldn't have any assurance that the commission had fully considered our petition, nor could we understand how the commission reached its conclusions, or even how we might remedy any shortcomings that the commission had identified.

Only a few days after we'd received the letter from the IACHR, the petition team received some sad news. Brian Tittemore, a lawyer working for the IACHR, had passed away from cancer on December 12, 2006. Brian had been very supportive of what we were doing, championing our cause from within, informing and educating the nine-man commission on what our petition was all about. Brian recognized what we were trying to do for the cause of humanity. His death was a great loss to our cause, but his passing was a great loss to the world as well.

At the same time that all this bad news was landing on my doorstep, I was also very sick with the flu. It was a low point for me—a time of confusion, anger, and loss of direction.

During a phone call with Martin, Donald, Paul, James Anaya, and the rest of the team, I acknowledged that hearing about the decision the way we did made the news even harder for me. I wasn't entirely surprised, given that there seemed to be a less-than-transparent connection between the State Department and the commission. But I felt the commission's response had been evasive and dismissive. "How could we possibly let the commission get away with informing the U.S. administration before informing us?" I asked in one of my emails. "And what kind of hold does the U.S. administration have on the commission?" Frankly, we felt the IACHR had let us down, as so many southern institutions had.

I went on:

In our attempt to be polite, respectful, not put the commission in the wrong light in order to keep the door open with them, I feel like I now have compromised something within me and all that I stand for. This process is very much about the journey being the destination, the power being in the attempt and the strength being in the struggle. It isn't only about the final ruling, per se, but how we Inuit portray ourselves and stand tall throughout the process, and how we as Inuit in turn are treated and respected throughout the process by institutions that are put there to protect our rights.

I also have looked up the commission website and saw for the first time they are all men from very warm countries. I do wonder if any one of those men [has] ever seen snow and sea ice, much less [has] any inkling as to what the ice, snow and cold represent for an entire people living at the top of the world.

James Anaya responded with compassion and empathy. He noted how hard IACHR lawyer Brian Tittemore had worked for our cause, and said he didn't know if his illness and passing had anything to do with the response we received. James's mention of Brian hit me hard. I had been focusing so much on how the commission had treated us that Brian's illness had been secondary to me. In fact, my denial had been so strong that when James mentioned that Brian had actually passed away, I was shocked. Brian was a truly good person, and I had always felt his support and encouragement. I remembered his openness and his willingness to help. All barriers gave way and I broke down and wept.

I AM A RATHER PERSISTENT PERSON. Despite the lack of response to my initial letter, I continued to write to the commission. I asked for a face-to-face meeting with them. I felt that it was crucial for them to hear us in person. In my letters, I insisted that they hold, at the very least, a hearing in Washington, D.C., on the legal impacts of climate change. Eventually, they agreed.

Paul Crowley, Martin Wagner, Donald Goldberg, and I took hold of this opportunity with full force, preparing for several days for the hearing. The commission would not hear the petition itself, but they would allow us to expand upon the legal basis for connecting climate change to human rights. They also asked if I would relate the impacts on other vulnerable regions of the world. Because I had always seen this issue in an inclusive way, I agreed to do that. That was what I'd always wanted.

When we got to the hearing in Washington, Paul, Martin, Don, and Dan Magraw from CIEL spoke to the commission about the various legal issues raised in our petition. In my speech, I addressed how climate change was affecting communities around the globe. I was careful not to try to represent the feelings and interests of other peoples, but explained the connections of the melting Arctic to other places in the world, and how rising sea levels were negatively affecting vulnerable regions, along with their Indigenous peoples. The month before I gave my speech, record-breaking winds in Iqaluit and Pannituuq had torn the roofs off buildings and homes. The weather, which we had learned and predicted for centuries, had become *uggianaqtuq*—a Nunavut term for behaving unexpectedly, or in an unfamiliar way. Our sea ice, which had allowed for safe travel for our hunters and provided a strong habitat for our marine mammals, was, and still is, deteriorating. I described what we had already so carefully documented in the petition: the human fatalities that had been caused by thinning ice, the animals that may face extinction, the crumbling coastlines, the communities that were having to relocate—in other words, the many ways that our rights to life, health, property, and a means of subsistence were being violated by a dramatically changing climate.

I also reminded them that global warming, which causes additional runoff from watersheds that empty into the Arctic, speeds the process by which POPs find their way into our marine mammals. And I reiterated that hunting for Inuit was so much more than a way of providing our communities with food. "Hunting," I said, "is, in reality, a powerful process where we prepare our young for the challenges and opportunities not only for survival on the land and ice but also for life itself. The character skills learned on the hunt, of patience, boldness, tenacity, focus, courage, sound judgment and wisdom, are very transferable to the modern world that has come so quickly to the Arctic world. We are seeing this powerful training ground on the land and ice being destroyed before our very eyes. Not only are our livelihoods being threatened but also we are losing lives as a result of these dramatic changes as the sea ice depletes and creates precarious situations for our hunters and their families."

I concluded by saying, "The individual rights of many are at stake. The collective rights of many peoples to their culture are also at stake. I encourage the commission to continue its work in protecting human rights. In so doing, you will protect the sentinels of climate change—the Indigenous peoples. By protecting the rights of those living sustainably in the Amazon basin or the rights of the Inuk hunter on the snow and ice, this commission will also be preserving the world's environmental early-warning system."

In response to my testimony, the president of the IACHR told me, through an interpreter, that they wanted to move forward with the petition. In fact, he told me that they had asked my legal team for some of the documentation, which was required for them to assess this further. He said that they would get back to us in two weeks. Martin, Don, Paul, Dan, and I headed home.

Those two weeks came and went without any of us hearing from the commission. In fact, as far as I knew, they never got back to us. It was a time of great frustration for me.

After my return to Iqaluit, I tried to find out if Earthjustice, CIEL, or any other environmental organization was working on some of the questions they had asked us to report back to them to assess the petition further. It was very difficult to get accurate answers. I wondered if the disconnect was because none of the legal team was attached to me in any official capacity now that I was no longer ICC chair. Yet I felt a great sense of responsibility to see the human rights work through, not only for myself but for my fellow petitioners. After the hearing, I sent a letter to each member of that group, updating them on the situation. But preparing sixty-two letters and having them translated into two languages proved to be a financial burden now that I didn't have any institutional support.

While our petition didn't get the reception we were hoping for, I believe it did play an important role in moving the climate change discussion forward. It wasn't evident immediately, but the petition was an important step in paving the way for other efforts to recognize climate change as a human rights issue. In fact, also in 2007, the Small Island Developing States met in Malé in the Maldives and, in partnership with CIEL, issued the Malé Declaration on the Human Dimension of Global Climate Change, which called on the United Nations High Commissioner for Human

Rights to conduct a study about how climate change affected the abilities of populations, like those in the sinking island states, to fully exercise their human rights. It also asked the United Nations Human Rights Council to convene a debate on climate change and human rights. Indeed, the following year, the council did adopt a resolution (7/23), referring to the United Nations' own charter and covenants on human rights, that acknowledged that climate change had "implications for the full enjoyment of human rights" for many peoples and communities around the globe. They reaffirmed this position in subsequent resolutions in 2009 and 2011. And in 2009 the UN High Commissioner for Human Rights did indeed release a study that looked at populations around the world whose economies, safety, and lives, as well as other human rights, were put at risk by climate disruptions. The United Nations continued to maintain its focus on the human rights angle at their COP 16 in Cancún. Along with calling for the establishment of a Green Climate Fund and a Climate Technology Centre and Network, the report produced by the conference echoed the earlier UN resolutions, demanding that climate change actions always take into account human rights.

NGOs outside of the environmental agencies have also brought the issue of climate change to their discussions of social, economic, and political struggles. In 2008 Oxfam International submitted to the UN High Commissioner for Human Rights a report entitled *Climate Wrongs and Human Rights*. In it Oxfam stated, "Climate change was first seen as a scientific problem, then an economic one. Now we must also see it as a matter of international justice. Human rights principles give an alternative to the view that everything from carbon to malnutrition can be priced, compared and traded."

And some communities have continued our legal pursuit of human rights, in a slightly different way. Kivalina, a small Inupiat community of Alaska, launched a lawsuit for damages caused by climate change against Exxon Mobil Corporation, eight other oil companies, fourteen power companies, and one coal company. Although the suit was thrown out in court, it helped to draw attention to the devastation caused by melting ice.

And another legal action connecting climate change to human rights has been brought to the Inter-American Commission on Human Rights. Four years after our legal impacts hearings, in March 2011, the IACHR heard testimony from Alivio Aruquipa, a farmer from the Bolivian Andes; Kevin E. Trenberth, head of the Climate Analysis Section of the National Center for Atmospheric Research; and Martin Wagner from Earthjustice, claiming that climate change was a human rights issue for Indigenous peoples in the Americas, specifically concerning water resources. The speakers stated that global warming had caused changing precipitation in the Americas and rapid melting of glaciers, changes that were threatening the peoples' ability to access the freshwater resources they needed for consumption, sanitation, and irrigation. As a result, livelihoods and the cultural survival of various ethnic groups in the regions were at risk.

The issue of access to water was presented to the Spanish-speaking IACHR by Spanish-speaking witnesses, and it's likely that the members of the commission were already familiar with the challenges that many people in warm climates have in accessing water. The hearing resulted in a press release from the IACHR that connected climate change to human rights. In part, it stated, "The Commission ... received alarming information on the already serious impact of anthropogenic climate change on the enjoyment of human rights, especially in mountain regions where the widespread loss of glaciers and snowpack and rising temperatures are diminishing access to water, harming food production, and introducing new diseases. The Commission urges States to keep human rights at the forefront of climate change negotiations, including in designing and implementing measures of mitigation and adaptation."

I realize now, with the luxury of hindsight, that there may have been a reason other than outstanding legal questions as to why our petition wasn't revisited, while the later hearing received such an endorsement from the IACHR. The concept of ice as the life force for an entire people at the top of the world was, I believe, too foreign to a commission made up of representatives from warm countries. It was difficult for them to grasp the fact that ice is something that people depend on not just for survival but to thrive. Indeed, the idea of "the right to be cold" is less relatable than "the right to water" for many people. This isn't meant to denigrate the people on the human rights commission and in the warmer countries, but rather to point out that the global connections we need to make in order to consider the world and its people as a whole are sometimes lacking. Because as hard as it is for many people to understand, for us Inuit, ice matters. *Ice is life.*

(There are two wonderful books that help to make clear the importance of ice to our people. *The Meaning of Ice: People and Sea Ice in Three Arctic Communities* is edited by Shari Fox Gearheard, Lene Kielsen Holm, Henry Huntington, Joe Mello Leavitt, Andrew R. Mahoney, Margaret Opie, Toku Oshima, and Joelie Sanguya and published by the International Polar Institute. *SIKU: Knowing Our Ice*, edited by S. Gearhead, I. Krupnik, G. Laidler, and L. Kielsen Holm [London: Springer], also explores this essential truth in moving detail.)

In the end, however, it's a positive sign that the "right to water" hearing succeeded the way it did. It provides further proof that people will continue to link climate change to human rights. Our Right to Be Cold petition may have been a little ahead of its time, but it seems to have been an important step in raising the world's awareness. Climate change is about people as much as it is about the earth, and the science, economics, and politics of our changing environment must always have a human face.

8

Acclaim from Outside, Peace from Within

I FELT A DEEP SENSE OF RELIEF upon leaving elected politics. After eleven physically and emotionally taxing years, I needed to slow down. I had been practically living on an airplane, traveling all over the world, and it was time for me to reclaim my sense of self and take care of my health.

I had been having some troubling physical issues during my last year with the ICC, but I'd been so busy that I hadn't had a chance to make a doctor's appointment when the symptoms first started to bother me. Now I was able to get checked and start the long series of tests that would eventually rule out an immediately serious condition. (I am closely checked annually through ultrasound for one condition, which according to the specialist is not the source of the discomfort I've continued to have over the years. I have learned to live with the symptoms and am no longer as afraid, which is what matters.)

While I knew I needed to attend to my physical self, I also wanted to regroup emotionally. Because I've had such a public profile for many years, most people don't realize that I'm actually an introvert—an introvert doing an extrovert's work.

This had been a real strain on me. I'd had to learn to overcome my fears of speaking in public and the discomfort I felt in large public events and crowded meetings. I always needed a great deal of personal time before and after these occasions to prepare and then recover. Years later I would read Susan Cain's *Quiet: The Power of Introverts in a World That Can't Stop Talking.* Her descriptions of introverts fit me to a T, and I was finally able to understand why I behaved and reacted in certain ways, that this wasn't some weakness or downfall in my character but, rather, a part of my psychological makeup. I have come to honor that in myself, and I relish my own company when I'm home. I live in complete silence most of the time. Back then, however, I just knew that I felt better in one-on-one conversations and when I had had plenty of solitude and time to spend on my inner life. And I recognized that eleven years of very public life, as well as the non-stop travel, had taken its toll. I was close to burnout when I decided not to run again for the ICC. I planned to stay home in Iqaluit, where I could embrace a newly quiet existence and turn toward the book I intended to start writing.

But this was not to be.

At seven in the morning on February 1, 2007, I received a call from Patricia Bell of CBC North Radio in Iqaluit. "Have you heard, Sheila, that you and Al Gore have been jointly nominated for the Nobel Peace Prize?" The phone call had woken me up, and I was still groggy. All I could get out was, "No, I'm still sleeping."

I discovered that I had been nominated by two Norwegian parliamentarians: former minister of environment and Conservative MP Børge Brende and Socialist Left Party MP Heidi Sørensen. I would also later find out that the announcement itself was unusual. The Nobel Committee does not disclose its nominations, but Brende and Sørensen, perhaps in a bid to draw the world's attention to climate change issues, had gone public with their nomination.

As I woke up to this news, quite literally, I realized that my hope of living a more quiet life had just been blown out of the sky. The phone rang constantly that day, and the next, and the next. The emails poured in, as did the requests for interviews and speaking engagements. After the disappointment of being told that the commission would not hear our petition, it was heartening to experience this outpouring of support from many of my colleagues and the institutions I'd worked with over the years, not to mention the people from our communities. And it was gratifying to see that the world at large did, apparently, want to hear about the petition.

The Nobel nomination wasn't the first time I had received accolades for my work on POPs and climate change. Just as my term as ICC chair was nearing its end, I'd received an unexpected call on my cell phone while I was in a cab, returning to the Ottawa office from a medical appointment. The voice on the other end introduced herself as a representative of the Sophie Prize of Norway. She informed me that the prize had just been awarded to me. It was not one that I was aware of, and in the confusion of the call, I mixed it up with the movie *Sophie's Choice*. I'm sure the caller was somewhat baffled by my polite but low-key response to her news. I would quickly learn, however, that the Sophie Prize was a real honor, and that I had been nominated by Lars Haltbrekken from Friends of the Earth Norway. Founded and funded by Jostein Gaarder, Norwegian author of the bestselling novel *Sophie's World*, and his wife, Siri Dannevig, the Sophie Prize was created to acknowledge those who are working to create a sustainable planet. It also comes with a hefty financial award that proved to be a lifesaver, helping to finance my continued work on the petition after I left the ICC and to give me time to focus on starting to write this book.

In the fall of 2005, I was awarded the Champions of the Earth Award by the United Nations Environment Programme. At around the same time, I was invited by President Bill Clinton to attend the inaugural opening of the Clinton Global Initiative (CGI) in New York City. Part of the Clinton Foundation, the initiative is a gathering of global leaders who are working toward solutions to the world's challenges. CGI's four areas of focus are poverty alleviation, religious tolerance, good governance, and climate change. In May 2006, I took up CGI's invitation to make a personal pledge to continue my efforts to put a human face on climate change.

While being invited to participate in CGI was an honor, it turned out to be not the most productive avenue for my work. The team that had put on the Arctic Wisdom event on the ice in Iqaluit in 2005 thought that we might do a second Arctic Wisdom gathering, this time in Greenland, and this time with the involvement of CGI and, we hoped, President Clinton himself. Unfortunately, this didn't pan out.

Those making pledges as part of the Clinton Global Initiative were expected to fundraise to support their commitment, and there were many rules to comply with as part of the process. With other work pressing in on me, as well as the high registration fee to attend the annual CGI event, I decided to focus on other ways of drawing attention to the human side of climate change.

While my involvement with CGI was relatively brief, it did have some wonderful highlights. One of these was meeting Hillary Clinton, for whom I have great respect and admiration. In 2004, she and Senator John McCain, flying above the fray of partisan politics, visited the Arctic in Alaska as part of a fact-finding mission on climate change. (In fact, both senators ended up appearing in the same video as I did, produced to supplement the Arctic Climate Impact Assessment.) During one of the CGI panel discussions, I asked Senator Clinton about implementing ACIA recommendations, and her answer assured me that she was aware of the assessment and understood the science. And during various CGI events, I also met fellow Sophie Prize winner Wangari Maathai of Kenya, who sadly has since passed away; environmental scientist Lester R. Brown; media mogul Ted Turner; and actor Brad Pitt. At the closing dinner, guests were even serenaded by the one and only Tony Bennett. Pretty big deal for an Inuk girl from the far reaches of the Arctic.

Interestingly, the one person I didn't meet during the CGI events was Al Gore. But I had heard him speak there, and the following year, as the petition team was preparing for the IACHR hearing in Washington, we decided to approach him. We called his office and asked if he would be willing to have coffee with me. His assistant told us that he was busy every day for the next six months. Now, however, with our joint nomination for what was arguably the biggest award in the world, I wondered if I would hear from him. After twenty-four hours had passed without a call, I realized that no connection beyond the nomination would be happening.

But while Al Gore didn't appear to be interested in making an Arctic connection through me, many others were. From the time the nominations were made public to the final announcement of the winners the following fall, I was deluged with media and speaking requests. Much of my previous work with the ICC had required appearances at international events outside of Canada. Now I was traveling around the world again, but also being asked to speak or give interviews in Canada. Indeed, the Nobel nomination seemed to make my own country wake up to my work, to ask who I was and what my message was all about in a way they hadn't before. And in June of that year the attention continued with my win of the Rachel Carson Prize in Norway and the Mahbub ul Haq Award for Human Development from the United Nations (presented by Secretary-General Ban Ki-moon). I received a number of other awards in Europe, the United States, and Canada, and was lauded as a "hero of the environment" and an "emissary" on issues of climate change by publications like *Time* and *Rolling Stone*. (In 2012, I would also be honored to have my life's work commemorated on a postage stamp, along with three great fellow Canadians: Louise Arbour, Michael J. Fox, and Rick Hansen.) Each award and acknowledgment was, to me, a symbol that my message was resonating with people.

I also received letters of support from around the globe for the human rights work we had been doing. One came from an elderly woman in Sacramento, California, thanking me for speaking up. She also included in her note a letter she had written to the U.S. Department of State, which said, in part,

> The message that Sheila and her Inuit people are sending out to the world is being heard and acted upon by ordinary Americans…. Even though my country has done nothing to curb pollution and degradation of the environment, we the people are taking action. Today, my carbon footprint is zero, and I live in a state where more and more local emission control laws are being passed. We can't do it alone, but it's a start. I'm an old lady on a fixed income with limited resources, and if I can make changes, why can't the United States government?

It was wonderful to know that my work mattered to so many.

At the same time as I was talking to the media and hearing from concerned people around the world, unbeknownst to me, letters of support for my Nobel nomination were being sent to the Nobel Committee by Inuit Tapiriit Kanatami, the national Inuit organization in Canada, and by the United Nations Permanent Forum on Indigenous Issues. When I was sent a copy of a published interview with the president of the Nobel Committee in which he noted that the committee was receiving letters of support for various candidates, but that these could end up being more detrimental than beneficial to a candidate's selection, I realized I was likely one of those he was speaking about. Yet these gestures of support from my fellow Inuit and the Indigenous community, as well as the nomination itself, had more meaning for me than winning the prize itself.

The announcement of the prizewinners was to be made on October 12. I didn't get much sleep the night before. It had been an exhausting and exhilarating eight months. International media, and indeed the eyes of the world, had been focused on the Arctic and on the story I had been trying to make known for so long. People were finally coming to understand the Arctic challenges and the plight of its people. For me, this was *the moment*.

That evening I sent an email to my legal counsel, Paul Crowley, who'd been with me through the thick and thin of the human rights work. "You know, Paul," I said, "regardless of the outcome tomorrow morning, this global awareness that is happening right now is, in fact, the win. It doesn't matter what tomorrow morning will bring. For me, this is the win."

The announcement came in the early morning. The Nobel Committee had decided to sever the joint nomination I shared with Al Gore, giving the Nobel Peace Prize to Al Gore and the Intergovernmental Panel on Climate Change, a group of hundreds of scientists from around the world who were working on issues of climate change.

The media, of course, wanted to hear my thoughts about the decision. So, early in the morning, after just a couple of hours' sleep and looking quite haggard, I went to the CBC station in Iqaluit and did a number of interviews on radio and television, including one for *CBC Newsworld*. There was, apparently, a great deal of disappointment around the world that the nomination had been split. But I wasn't upset. Rather, I recognized that there were many other wins that day. The planet itself was the real winner with this award, and so it was a victory for all of us. Climate change champions received the prize, after all, so it was truly a win for the Arctic and for us Inuit. And if getting the United States on side was our ultimate goal, having Al Gore, a well-known American, win the award was part of a good strategy on the part of the Nobel Committee.

The only disappointment I felt upon not winning the Nobel Peace Prize that morning came from the sense that I had let people down. I'd been receiving such extraordinary support from my Inuit community that I really felt we would all be winning this together. And I knew many Inuit were counting on the win to help boost our collective sense of pride, confidence, and self-worth.

After I had finished my interviews in the studio, I returned home. Although I hadn't won the prize, a palpable sense of calm came over me that quiet Arctic morning. In a sense, this feeling was a win for me as well. All the relentless media work and constant public engagement was finally over—or so I hoped. The work I had been doing on the environment and climate change had led me to become a global citizen. For many years,

I had, in a way, belonged to the world in my various roles, and that had taken a lot from me as an individual. If I had won, I would have *truly* belonged to the world. Emotionally and physically, I may not have been able to deal with the publicity: it certainly would have taken more energy than I felt I had at that time. Perhaps I just wasn't ready for it. As I drank a cup of tea and gazed out my living-room window, across the icy waters of Frobisher Bay that stretched as far as the eye could see, I realized that my non-win had given me a true sense of inner peace.

PEOPLE OFTEN ASK ME how I got to where I am, how I managed, from such humble beginnings, to reach a point of global recognition, honorary degrees, and international awards. Of course, the work I did and continue to do, and the deep sense of responsibility I feel toward my Inuit community, as an elected official and as a mother, grandmother, friend, and neighbor, drove me forward. But there has also been a deeply personal component to all this. During the Arnait Nipingit Women's Leadership Summit in 2010, I was asked what leadership means to me. "Leadership," I told the group, "means never losing sight of the fact that the issues at hand are so much bigger than you. Leadership is about working from a principled and ethical place within yourself. It is to model, authentically, for others, a sense of calm, clarity and focus. Leadership is to always check inward, to ensure you are leading from a position of strength, not fear or victimhood, so you do not project your own limitations to those you are modelling possibilities for." That "checking inward" and the personal growth that accompanies such introspection have been, I believe, instrumental to my own ability to succeed.

Many of the points of recognition, accomplishment, and progress in my career came at moments in my life when I had worked through personal issues and discovered some truth that allowed me to stick to my convictions and commitment, to focus on the goal and not the obstacles. And each time I decided to stay the course, providence seemed to take over. As the saying goes, "When a decision is made, the universe conspires to make it happen." I believe that all of us have what it takes to move toward healing, work through our issues and woundings, rewrite those histories, and change our perspectives. That kind of inner work gets reflected in the outside world. For me, the accolades and awards I've received are really about my inner mirroring.

I like to think of our hardships, baggage, or woundings as "holes in our souls," and I believe that if we don't close these holes, we pass them down in some fashion to our children. In this way, the historical traumas of our Inuit elders have been passed down to the younger generation. But I also believe that the many tests we face in life often arrive to help us mend these gaps in our souls. One of the holes for me was my fatherlessness.

While I never felt, when I was a small child, that our family was lacking in any way, the years in Blanche and after, and the continuing questions about my white look and my parentage, made me begin to feel that something was missing in me. That insecurity followed me well into adulthood. (Sadly, I still get these questions. A few years ago, a Canadian journalist who knew me quite well blurted out, "It's too bad you are so damn white-looking." I think she was suggesting that my Arctic message might have greater impact if I were more Inuk-looking, but her comment stung. I wish I'd had the presence of mind at that moment to respond the way some young Inuit of the Nunavut Sivuniksavut program recently responded at a public event in Ottawa when asked about their pale complexions. "This is how many Inuit look today. Get used to it.")

When my daughter, Sylvia, was a teenager, she suggested that I do something to fill this hole in my life and actually look for the man whose name was on the certificate that Uncle Jamie had given me when I headed to Churchill. It was, of course, a wise suggestion. My brother Charlie had connected with his father years before and had had the opportunity to get to know his half-siblings. My adopted brother, Elijah, had also done the same thing a long time ago. Bridget, sadly, never had the chance to meet her Irish father before she died.

Someone, perhaps Uncle Jamie, had mentioned to me that George Kornelson had gone out West after his time in the Arctic. Following Sylvia's suggestion, whenever I traveled out West, I would pick up a phone book and look for his surname. During one visit to Calgary, I called a Brian Kornelson, who was listed in the book. I pretended to be gathering information on the history of Arctic Quebec. I told him I knew that a George Kornelson had worked in the North in the 1950s, and I asked if Brian knew him. Brian replied that this must have been his great-uncle, but that he had died ten years ago. I thanked Brian and shakily hung up the phone. In the hotel room in Calgary, by myself, I grieved the death of a father I had never known.

Many years later, in 1991, while I was working at the Makivik Corporation in Montreal, Martha Montour, a Mohawk woman I'd shared a podium with at a college talk, introduced me to an Aboriginal RCMP constable named Jim Potts. I told him about my search for my father and asked him for help in getting his files from the RCMP. He agreed almost immediately. Within two weeks, I received a folder of information that included a picture of my father and an account of his postings. There was a letter stating that in 1954, the Anglican minister at Fort Chimo, Jamie Clark, had informed the RCMP headquarters that he'd baptized the illegitimate child of Daisy Watt, and that she had stated Kornelson was the father. The letter continued that Kornelson had said that he had never been physically intimate with Watt. George Kornelson had stated that my mother already had two children by two different white men, implying that my mother's statement that he was the father of her third child was questionable.

Jim Potts wrote a note at the end of the file to help contextualize my father's files for me:

> This information is not overly complete, but I hope you find it useful. In a historical perspective, in the 1960s and before, the RCMP were very paternal towards its members. Members were discouraged from marrying native women. If a member became involved, he was usually transferred on short notice. It's safe to say that if Kornelson had admitted to being your father, he may have found it difficult within his new family and within the force. Believe your mother.
>
> Best of luck,
> Jim Potts

Initially, I reacted negatively to the information in the file. I focused on the fact that George Kornelson had "invisibilized" me and my existence through his denial of his relationship with my mother. And I lingered on all the damage he had done in my family. For I knew that not only had he left my mother but he had also betrayed her in the worst way. When my mother was pregnant with me, she discovered that my father had seduced her sister, Penina, who had become pregnant as well. My mother felt a deep sense of betrayal, and both women were devastated by what had happened. My mother turned her anger outward. She wanted nothing to do with my father when she discovered what he had done. But Penina turned her pain inward and remained marked for most of her life. Her son, Elijah, was born six weeks after me, and while Penina eventually married, her husband had difficulty accepting Elijah as his stepson. Like his mother before him, young Elijah ended up being raised by the Shipaluks. While the Shipaluks were a loving family, Elijah has had enormous struggles for most of his adult life but is now finding a better balance. And now, adding to all this hurt, I discovered that George Kornelson would not even acknowledge his relationship to us.

Despite my continued success in my career, with these thoughts weighing on me, my personal life spiraled downward for a number of years. It wasn't until about eight years later that I ended up looking at these documents again, this time through different eyes. As I reread them, I saw in them not a sad story of fatherlessness, denial, abandonment, and rejection, but something else. I realized that while my father had denied my family's existence, my uncle Jamie had always been present. He was there from the beginning for me, and he made sure that the record stated that George Kornelson was my father, that there was a document acknowledging that I *had* a father.

I immediately got on the phone and called Uncle Jamie. As Aunt Ruth passed the phone to him, I heard his familiar chuckle. "Is that *really* you?" he said.

I could feel my heart open. Those exact words, "Is that really you?" and that voice were really what I had always longed to hear if ever I found my father.

After the death of my sister, Bridget, I wrote Uncle Jamie a letter. I wanted to tell him how I had come to the point of making that phone call, and how much I had needed him at that time.

November 16, 1999

Dear Uncle Jamie,

It was wonderful to hear your voice and catch up on some news. I wondered why I waited so long to hear that soothing familiar chuckle once again....

Although I've always had great faith, I have chosen in these last 10 years to deepen my spirituality, and have it reflect in the work that I do. I have worked and prayed hard to come to terms with the challenges in my life, and have come to fully understand that everything happens for good reason.

In other words, the not-so-perfect childhood that I was given was part of a perfect plan for God to prepare me for what I was to become, and what I was to do with my life. The soul search has led me to many levels of consciousness, awareness and clarity that, indeed, we are being led in the best, as well as the worst, of times, which brings me to the coincidence of having your nassak [hat] and your Zippo lighter find their way back into my hands, and how that has deepened further my faith, and that I am indeed in the best of gentle hands....

Throughout these last months since Bridget's death, I have sat in a void, deeply feeling the loss of her supportive gentleness, a gentleness that I unknowingly had taken for granted until it was gone. I wondered how I would do without it. Then the nassak and the Zippo lighter of yet another significant gentle spirit of my childhood found their way into my hands, brought by yet another gentle spirit, starting a new life in the Arctic....

Last summer in July, I was invited to speak at an event in Iqaluit and stayed at a breathtakingly beautiful new bed and breakfast. I had met many people there that week, since it was a full house, and one of the people there was a pleasant, soft-spoken man, who had just arrived in the Arctic from London, Ontario. He was hired to work for the Nunavut government and is stationed in Coppermine. However, his work brings him frequently to Iqaluit. A pleasant connection was made with him, and as I parted back to Ottawa, I wished him well in his new work and said, "The North is a small world. Perhaps we will meet again." I went back to Iqaluit for more meetings in October, and again, he was there at the bed and breakfast, doing more work in Iqaluit. The connection deepened, and we talked late into the night, and during the course of our conversation, he showed me a nassak and lighter that he had bought at a garage sale in London from an Anglican minister who had been in Nunavik for many years. Of course, your image and energy surfaced immediately, and I could barely gather my thoughts and feelings as I held your nassak in almost dumbfounded shock.

You see, Uncle Jamie, I knew immediately it was yours.... I'd been on this journey long enough not to interpret coincidences or what I call God's messages too quickly, and I left back for Ottawa, having trust and faith that this was yet another event leading to the deepening of my spirituality.....

This is a long story, I know, but you see, Uncle Jamie, I needed to share it with you and thank you from the bottom of my heart for giving me gentle support throughout my life. As I said earlier, it's not only harshness that has been in my life, but it is almost as though, now that the dust has settled, I am able to see and truly appreciate the significance of this gentle love and kindness that I received from you and Bridget. I also believe that it now allows for new space to be created to have ongoing love and gentleness in my life. I feel I am at some turning point....

God bless,
Sheila

When Uncle Jamie died in 2006, I went to his memorial and spoke about his role in my life. I had come to know, through Uncle Jamie's love, that the universe had given me the father spirit I needed. Indeed, I realized then that the universe, or God, gives you what you need, the people who can fill the voids in your life, but it's up to you to recognize them. You have to empty yourself of pain and angst before you can recognize the love and warmth of another. It's when we see what the universe, or God, however we describe our higher power, has provided for us that we learn how to fill the holes in our lives. It's how we become complete within ourselves, in a sense.

Even before my realization about the role Uncle Jamie had played in my life, I had a number of "aha" moments that helped me through difficult times. One was during my trip to Nairobi for the second negotiating session on the POPs treaty. My friend Gail had given me a book by Gary Zukav, called *The Seat of the Soul*. I brought it with me to read during the long plane ride down to Kenya and during any breaks from the meetings. It was nice to focus on something other than work. The ICC had just finished its Inuit Express mission to Russia, and in answer to a critical assessment in *The Globe and Mail*, I had written an editorial. Unfortunately, the head of ICC Greenland didn't like the approach I'd taken, and things were tense between us. It wasn't the first time I had butted heads with this colleague, and I was feeling drained by all the drama. So when I got to the chapter in Zukav's book entitled "Intention I," I was stunned by the following passage:

> A nation is an aspect of the personality of Gaia, the earth's soul, which itself is developing its personality and "soulhood." The group dynamic that is the United States is a personality aspect of Gaia, as is the group dynamic that is Canada, and the group dynamic that is Greenland, and the group dynamic that is each nation. The individual souls that participate in the evolution of these aspects of Gaia form these group energy dynamics, and at the time, their own group developments are served by the karmic energy properties of these nations.

I couldn't believe what I was reading. I shut the book, dumbfounded. *Greenland?* Who knows of Greenland, much less writes about it in terms of the earth's group dynamics, group energies? Why not Denmark or Iceland, countries that are better known to the world? Why the United States, Canada, and Greenland as the examples? Those are the three countries where we Inuit live! Sure, America and Canada are close to each other, but Greenland?

Only moments before I opened the book I had been praying for the personal crisis that the conflict with my ICC Greenland colleague had brought about to be lifted from me. The coincidence produced a powerful reaction in me. I couldn't form a rational explanation for what I had read, so instead I allowed my heart and soul to fully feel the miracle of the moment. As I sat in that beautiful Safari Park hotel room, far away from my homeland, I wept. I let my emotions wash over me. I realized then that there really is a cosmic order to life. I had never felt such deep trust and faith as I did in that moment. In fact, I call it my "holy moment."

Since my experience reading *The Seat of the Soul* in Kenya, I have come to understand that God is very precise. There is a space and a place in the world's unconsciousness, in the universe, where all things are known, where all things are understood, and where all things connect. We are always given signs if we stay alert to them, which in turn allows us to see that our personal experiences and our interpretation of them can have great meaning in the larger scheme of our life's journey.

I didn't realize until later that I'd been given that powerful moment to help me survive the sudden death of my sister, which happened six days after I returned home from Nairobi. Later, Gary Zukav's work continued to help me to better understand and accept that spirit chooses to come and spirit chooses to leave in its time. This wisdom helped me to not only focus on the "death day" of my sister and other deceased loved ones but also move beyond it and accept the remarkable role that these family members played in my life when they were alive.

Later on, I wrote and aired a piece for a CBC Radio program called *This I Believe*, detailing this time in my life. In it, I said,

I believe that how we deal with the losses in our lives can help us realize our potential as human beings. Everything that happens to us is interconnected—with other people, with nature and with each stage of our journey through life.... My life-defining and greatest personal test started in February 1999, when my beloved and only sister died suddenly of a massive heart attack at the age of forty-eight. My entire world stopped. For a very long time I could not let her go. Over the next five consecutive years, still grieving deeply over her death, life continued to test my resolve. I lost four more close family members—my aunt, my mother, a young cousin and a young niece. I thought I would never stop grieving. I wept literally for years—in foreign countries, in strange hotels, in airports—whenever the waves of sadness would overcome me.

In those places of deep grief, I deepened my personal journey. Each loss held me in a place I needed to be until I gained insight and clarity. I could eventually translate my new perspectives into powerful opportunities for personal change and growth. I came to see in a vivid way that all things are interconnected and that all things happen for a greater cause. I came to know trust in the life process. I came to know courage, tenacity and commitment. I needed these character skills in order to survive my grief. As it turned out, I also needed them to strengthen and raise the volume of my own voice on the global stage. There were many times I thought I could not carry on. But I learned that true commitment really begins when we reach a point of not knowing how we can possibly go on—and then somehow find the strength to go on anyway.

I used my commitment and the potent lessons I

learned to pioneer the work of connecting climate change to human rights on the global stage. I now honour my losses by turning the energy of dealing with loss to protection—protection of this wise hunting culture upon which I began my life with my small and close-knit family. And trust me when I say, this work requires much trust, boldness, courage and commitment.

Throughout the time of these losses and in years to come, I continued to struggle with my father issues. Indeed, I came to realize that I was grieving not just my own losses and traumas but also those of my mother, my aunt Penina, and my grandmother. I believe if our parents and grandparents haven't had the opportunity to deal with their issues, their children carry that baggage and have to work through it for them. My grandmother had never resolved her feelings about her abandonment by my grandfather. Neither had my mother fully worked through the pain and hurt of having the fathers of her children leave each time. I came to feel that my own healing was helping my ancestors reconcile, that they as women were working through me.

Part of this process happened when I was invited to receive an honorary degree at Bowdoin College in Brunswick, Maine, in 2008. I have always felt that I was being led to certain places for a reason, and that if I followed, I would learn the right lessons and gain insights that would strengthen the essence of who I really was. So if I was invited somewhere, whether for an event or for employment, I would know that I was going there to further my life's work, but also to work through issues on a spiritual level.

During the ceremony at Bowdoin College, we received our honorary degrees outside, in a large tented area. I'd brought

along two friends, Ann Vick Westgate, who had been part of
the Nunavik Education Task Force work, and her husband,
Michael Westgate. After the ceremony, they asked me if I had
seen the eagle that had been flying above the tent. I had not,
but my mind immediately raced to a piece that had arrived in
my mailbox in Iqaluit a year earlier. In a 2007 newsletter from
the National Aboriginal Policing Forum, the chair wrote:

> During the summer of 1999, I had the good fortune
> of attending a sacred camp on a reserve in southern
> Saskatchewan. Some 75 people from across Canada,
> mostly non-native, gathered for four days to take part
> in ceremonies and hear traditional teachings of the local
> native people. On the morning of day one, our first task
> was to raise three large tepees. Many pitched in, and after
> about 90 minutes, the final poles were being put in place.
> Inside the last tepee, while driving the stake to secure
> the tie down rope, the elder I was helping pointed to the
> opening at the top, and said, "Now watch, the eagle will
> come." I watched for perhaps two minutes, when out of
> nowhere, an eagle suddenly appeared and began circling
> high above. I turned to the elder, and asked, "How did
> you know?" He only smiled and said, "Now that the
> eagle has come, this camp will be good." And it was. Like
> this sacred camp, people attending this conference will be
> from across Canada. Come and share your wisdom and
> teachings with us. Perhaps the eagle too will bless us with
> his presence.

That chair was Jim Potts, the very same inspector who'd
given me my father's files from the RCMP. I honor the fact
that the eagle represents many things to many people, but

to me, the eagle that flew over the tent at Bowdoin was a connection to my father. I thought I'd already come to terms with the issues surrounding my father, but clearly my visit to Bowdoin was indicating that there was more to be done in this place.

A year after the convocation, I was invited to Bowdoin as a visiting scholar to teach the human dimension to climate change, along with Susan Kaplan of the Peary–MacMillan Arctic Museum. As I was driving down to Brunswick, where I was to live for the next year, I was overcome with feelings of doubt and apprehension. I arrived in the middle of an August heat wave. It was my first time teaching in a university setting, and being far from home, in another country, without the comfort of my precious country food, I had a trying and somewhat lonely time. Sharing lunches with Kristi Clifford, who worked for Susan Kaplan and became a fast friend, and suppers from time to time at the home of fellow-Canadian Genny Lemoine, who was curator of the museum, provided companionship and were an enormous help in getting through the year.

But after the first semester was over, I had to admit that while I had enjoyed the remarkable, open-minded American students, I didn't feel that I'd been able to give my whole self to the class. In other words, I felt I had to "put aside" a large part of myself to conform to the needs and expectations of the classroom. The constant angst I was feeling led me to further reflect on the journey I was taking, and I soon realized that I needed to address my Scottish grandfather, the first "severing" of father that had taken place in my family. For the first time since I'd arrived, I opened up a trunk I had brought with me. As I sorted through the contents, preparing to resume the writing of this very book, I came across a letter that my grandfather had sent to my grandmother after he left her in Fort Chimo with two young children and one unborn. I hadn't known that this letter existed until my mother's death, when we went through her personal effects, and until now I hadn't thought more deeply about it.

I held it in my hands and began to read the words that my grandfather had penned decades ago. It was written on Hudson's Bay Company letterhead.

My dear Jeannie,

How are you getting on since I left Chimo? I've been thinking of you often since I left. In fact, I'm sorry you did not come along with me, for you would have had a good time. It's not half as bad as I thought it would have been here.

They are getting ready to start, so I will say goodbye. Stay in the house, and never mind what people say. They're always talking anyway.

Kind regards to Emily Okpik, and all the Eskimo around Chimo. I'll write again by the steamer.

Yours truly,
William H. Watt

I hadn't looked at this letter for almost a decade, and now I read it over and over for a couple of days, just trying to comprehend the meaning behind all this. All these years, I had been trying to get to the bottom of my father issues, yet this really meant letting go of my whole family history of abandonment, and forgiving not only my father, but also my grandfather.

I felt as if, all this time, my grandfather had been there

waiting to say, "How about me? How about forgiving me? I was the first to leave your family. I was the beginning of the woundings that have been passed down to you."

It also made perfect sense to me that I would be at Bowdoin when I would come to terms with this intergenerational history of fatherlessness, that, so far from home, I would be grieving, weeping, cleansing myself of the old energies that I no longer needed. Since my divorce, I had often set myself on a path of rejection, as I was drawn to emotionally unavailable men. (That's who fatherless girls always want to pursue, because what else do we know?)

These emotionally distant, elusive men clearly reflected the absenteeism that was all I knew of my father and grandfather. Now here I was in a place that embodied that distance, elusiveness, and foreignness. That self-knowledge brought me an emotional, spiritual peace. I felt I was finally finding the true essence of Siila. My healing and wholeness as a woman were coming together.

Although learning of my father's death years earlier had been a sad postscript to a story of loss, there were also remarkable closures. In 2007, unbeknownst to me, my daughter had found my father's other children—my half-sister and half-brother—through Facebook. Soon after, through a synchronous chain of events, Sylvia was at my half-sister Susan's place in St. Catharines, Ontario, catching up on the connections. Sylvia arrived home a few days later, the day before Father's Day, bearing gifts from Susan. There was a lovely card and a photograph of my father. Unlike the picture of a stoic figure in uniform I'd received from the RCMP, this was a shot of a smiling man. In another bag were his RCMP boot spurs. Tears came to my eyes as I held something of my father's for the first time. I am forever grateful to Susan for being so open to accepting me as the sister she said she'd always longed for. Although I met Susan twice, I was only able to fully express that gratitude two years later at her memorial after her sudden death. In my tribute to Susan, which was read by my half-brother, Alan, whom I had yet to meet at that time, I thanked her for being so open to accepting me and for bringing some closure to my "invisibility" issue, which our father had helped to create by denying a connection to me.

After my year at Bowdoin College, I went to co-teach with Ian Mauro at Mount Allison University in Sackville, New Brunswick, for a year. I drove from Sackville to Charlottetown, Prince Edward Island, symbolically crossing the longest bridge in Canada to spend the Thanksgiving weekend with my half-brother and his wife, Darlene. It was wonderful to have Alan cook for me, and I felt at ease in his company.

I learned more about our father from Alan, a kind man who was pleased to meet me, especially in light of having just lost his only other sibling. I had now come full circle with my Kornelson family, as I'd met my uncle Ed, the only surviving sibling of my father, in Kelowna, British Columbia, the previous spring. He too was a lovely man who accepted me fully as his brother's daughter. It was, in fact, like meeting my father. This had been a long journey for me, but I suspect it had its purpose and meaning and connection to the exact challenges and struggles I needed in this lifetime. My hope is that those reading this will find resonance and somehow see themselves in my story, and know that we can, if we pay attention to life's signs and the lessons our struggle gives us, come to terms with our legacies. For Inuit, I truly believe the full meaning of our lives and the reason we are all here is to rewrite our histories.

I think that often the success of my work—the accolades, the rewards, the recognition—taken altogether looks Herculean. People don't realize, however, that it took many steps to get here. Part of the process was working on these holes in my life, on my personal healing. As an elected official, I always tried to make sure that I wasn't working through some emotional dysfunction of my own and projecting it in my work. I never wanted to overreact to something that I didn't need to focus on. For my work to excel, I needed to remain solid within myself so that I could model clarity and wisdom, rather than fear.

To put it another way, I see my life as being made up not just of political wins, but equally of spiritual wins. The obstacles that I overcame were reflected in the world. If you interpret your life that way, then your ego can't balloon, because you're simply humbled by the process. Your success is put into context. Looking back at my work, as public as it has been, I have to say that it has been, above all, a deeply personal journey.

9

Citizens of the World

In 2012 I attended the International Polar Year Conference, where the keynote speaker was ICC chair Aqqaluk Lynge. Aqqaluk has been involved with the ICC since day one and has always been thoroughly committed to what the council stands for. In his remarks to the conference, he explained the organization's history and political strength, saying, "We are not isolationists. We want to work with people from the outside, especially with scientists, because I think we have common goals." But he also pointed out that "the creation of ICC came out of the realization that Inuit needed to speak with a united voice on issues of common concern with regards to resource development and political actions in the Arctic."

During my time with the ICC, we directed a great deal of our focus and energy toward the POPs crisis and addressing climate change through human rights. But Aqqaluk was right— resource development has always been a central concern for the organization. The Arctic is, after all, a territory rich in natural gas, oil, and minerals. For many years, various corporations have been pushing to expand operations in the North to extract these minerals and fossil fuels from beneath the ice. With the sea ice and permafrost of the Arctic rapidly melting, these mining operations are becoming more and more feasible and potentially profitable. And these days, many in our communities feel that this may be our best path to future economic growth and independence. In particular, the governments of Nunavut and Greenland are now moving strongly in the direction of development, including oil and gas extraction.

It's understandable, given the poverty, lack of food security, and increasing difficulty of maintaining our traditional hunting culture, that the lure of resource-related jobs would be so great. But mining means digging up the land that we hold sacred, the land that we Inuit have been profoundly connected to for our entire history. Just as important, resource extraction poses serious environmental risks. All over the world, Indigenous peoples have suffered the devastating effects of these industries on their lands. One need only look to the constant oil spills in the Niger Delta or to the attacks against Peruvian Indigenous villages from the mining sector or to the effects of BP's oil spill on Gulf of Mexico fisheries to understand that resource extraction is, in large part, a dangerous and intrusive industry.

In many parts of the world, people are, indeed, fighting against the considerable damage done by resource industries. In 2011 Ecuadoreans won a case against Chevron Corporation in an eighteen-billion-dollar oil-pollution lawsuit that was upheld by an intermediate appeals court in Ecuador. In 2012 they continued the case by filing an enforcement-action lawsuit in the Superior Court of Justice in Ontario, seeking the seizure of shares and assets of the corporation. While this case is ongoing, the South Americans have attempted to reach out to our Inuit communities in Greenland to warn them of the devastation that the company caused to their people in Ecuador.

Of course, battles between the resource sector and the people who live on the land are happening in our own country, too. Today, the Dene, Cree, and Métis communities are the most negatively affected by tar sands development. For the Fort Chipewyan communities of Alberta, the contamination of the Slave River has affected their ability to fish, hunt wild game, and access a source of potable fresh water. Despite findings of elevated arsenic levels in the river, increased rates of cancer between 1995 and 2006, and noticeably deformed fish found in the river, Fort Chipewyan communities have continually struggled to prove to the Canadian government the connection between tar sands development and the community's diminishing health.

In an email, scientist David Schindler commented to me about the extensive damage of the Canadian tar sands, saying, "To me, tar sands are bad because of the total package. Greenhouse gas emissions, water use, water pollution, destruction of habitat for fish and animals, violation of treaty rights, the threat of tailing pond leaks. The total package is what makes them unredeemable development."

In the chemical valley along the St. Clair River in the Great Lakes basin, the Walpole Island and Aamjiwnaang First Nations communities have suffered as a result of industrial pollution upstream. After a number of chemical spills, these communities have had to adjust their diets and, at times, completely eliminate food and water sourced from the river. The residents have noticed various health effects along with these forced cultural changes, including increased birth defects, shockingly low rates of live male births, and high rates of cancer.

Given the wealth of evidence about the dangers of the extraction industries, we Inuit should be asking ourselves, "Why would it be any different for us? How will this industry, which is so counter to our own culture of stewardship of the land, be any different in the Arctic than it has been in other parts of the world?" If social and financial progress is a goal for our Arctic communities, we must seriously question whether we, as Inuit, are really and truly going to gain long term from resource extraction in the North.

And yet a number of our own organizations have started to embrace a resource-based economy in the Arctic in a serious way. Some communities are already pushing through a reformed educational curriculum to train Inuit children, as early as grade three, to work in the mining sector. As of April 2012, Agnico Eagle Mines Limited and the Nunavut education department have signed a memorandum of understanding, in which the mining company agreed to contribute four hundred thousand dollars a year over three years toward in-school programs aimed at training our youth for future jobs with Agnico Eagle. They are already working on a Nunavut-specific "Mining Matters" science course for elementary school children, and a work-shadowing program for high school students studying trades at Agnico Eagle's Meadowbank and Rankin Inlet Meliadine gold mine projects. In other words, they are tracking children to ensure that they pursue careers in that industry.

Beyond the commitment to a mining economy, which is troubling in itself if that is the only option being offered, this educational approach has disturbing colonial echoes of the streaming of immigrant, black, and Aboriginal children into vocational programs on the assumption that they "couldn't do" academic work. It certainly will have the effect of brainwashing many of our youth into going into one specific employment sector. And it limits the creativity, critical thinking, and innovation of our Inuit children, innovation that has in the past given us the *qajaq* (kayak), the *illuvigait* (igloo), and all the other ingenuity that allowed our people to thrive in the world's harshest environment. It ignores the need to more effectively include our remarkable, wise Inuit culture in our educational programs, and instead focuses young minds on a narrow set of job-related skills. I understand that this approach is coming from well-intentioned leaders who may feel as if they have no choice but to take this narrowly focused direction due to lack of funding and support from the federal government for our Aboriginal communities' educational needs. (I also appreciate that this is not the only route they are taking to better our education system, despite their limited funds.) If our federal government would contribute equal or higher amounts than the mining companies are offering, our Inuit children could be given many other options to choose from as they find their own life's passion and purpose.

While our Inuit world has made much headway in tackling the challenges that have come with modern ways, after all this time, we're still facing poverty, social and health problems, and many other struggles because we, along with our governments, haven't learned how to empower our communities to become prosperous and sustainable regions. Faced with this dilemma, we're putting all our eggs in one basket, assuming that development will be the solution. There's great irony in trying to solve these social problems by allowing the extractive industries, so responsible for social and environmental problems in nearly every area of the world, to come into our already-struggling Arctic communities. And because we have moved from a traditional way of life to a new modern world in a short period of time, are we not perhaps even more vulnerable than some of the areas that have already been affected by these industries? We have barely managed to deal with less intrusive institutions that have come into the North over the last five decades. We're still trying to pull ourselves out of dysfunctional, toxic relationships with many of these institutions that are still not really our own, as they are not based on the foundation of our wisdom, language, or culture. How is it that the extraction industry is going to work better for us? This is one heck of a risky business we're getting into as a means of pulling ourselves out of poverty.

And even if we come to the conclusion that developing our resource sector is the only way to provide financial security for our communities, will this really work? The extraction industries have a history of failure in creating a sustainable economic boost in the territories they occupy. First of all, the extraction projects are temporary and bring in an influx of money for only a short period of time. They are driven by global markets, and if the demand for the resource drops, so will the companies' interest in the Arctic. (Will the skills that Inuit learn on these jobs be transferable to other work once the companies pull out of the Arctic?) What's more, the companies often bring in temporary labor from outside the area for many of the jobs, putting pressure on local housing stocks and services and sending money back out of the local communities. In fact, all over the world, these intrusive industries have, in almost every case, caused greater economic and social hardship on communities than any gains from extracting the rich resources. I ask the question again, why would it be any different for us in the Arctic?

In the early 1990s, I joined Inuit Tapiriit Kanatami and other ICC elected leaders and staff on a trip to Harvard University to take part in a leadership-training program. One of Harvard's studies on economic development was a real eye-opener. Professor Joseph Kalt of Harvard had just finished a study on Aboriginal reservations across the United States, examining variations in approaches to economic development. Although the research was complex and extensive, one of the things that stood out to me from his findings was that no matter how rich the resources were around certain reservations, if there wasn't a "culture match" between the community and the actual work itself or how the business was conducted, these economic endeavors failed. However, if the values and principles of the particular tribe were honored and respected, and everyone in the community had a sense of control over this new development, including the work plans, the policies, and the manner in which the business was developed and structured, success was more likely. These findings became part of the recommendations we made for our institution in the Nunavit Education Task Force review all those years ago. I find myself revisiting this study when thinking about the resource-extraction industry's movement into the Arctic.

Finn Lynge, a remarkable, accomplished academic Greenlandic Inuk who contributed greatly to our Inuit world and who has recently passed away, identified four values that are essential to the Inuit way of life. *Nunamut ataqqinninneq* relates to a sense of pride and respect in a strong familiarity with and knowledge of the land and sea, including its animals. *Akisussaassuseq* is the responsibility that people have to the land and to everything that inhabits it. *Tukkussuseq* relates to the importance of generosity and hospitality to extended family, as embodied by the cultural sharing of our hunt. *Inuk nammineq* emphasizes individual autonomy and strength, particularly the wisdom to discern appropriate choices on the land and in one's life.

It seems unlikely that the resource industry could honor that second value, *akisussaassuseq*, when they are in the business of drilling, excavating, and disrupting the very land that we have held sacred for millennia and with which we hold a deep connection. How will our hunters, men, women, and youth, those who have known the wisdom of the land, feel at the end of a workday spent digging up and destroying the very land they have held sacred? Will these short-term jobs really address the dispiritedness of our men, a dispiritedness that is at the root of the social and health ailments of our communities? In other words, will the resource industry help make our communities safer? Will it help make our women safer? Will it help make our children safer? Have self-destruction and suicide rates been lowered in other parts of the Indigenous world where resource extraction has been used as the means to pull communities out of poverty?

I believe that many of these industries, unless they reinvent themselves entirely, have the potential to cause further disconnect and severance from our unique and remarkable culture, perhaps in even more ways than past social, educational, or judicial institutions that have come into our world. Over the decades, the speed at which modernization and globalization hit our world negatively affected our ability to remain steadfast in our values. The resulting loss of our principles has led to monumental dependencies and has eroded our identities and sense of self-worth. We are still trying to deal with the aftermath of those institutions and figure out a culture match with their new incarnations. What will mining, oil and gas, and, in fact, all extractive industries do to our already-vulnerable collective hunter spirit? Will the industry adapt itself to our needs, or will it actually cause more harm to our Inuit culture?

It's important to note that my stance on these issues is not intended to demean or demonize those who lead and work in these industries, any more than I demonized those

I struggled with in my past regional education work and in my international work as I launched the legal petition targeting the United States for their inaction in addressing their CO_2 emissions. In fact, there are Inuit who are seizing the opportunities to develop their own businesses around these industries in our Arctic. My voice on these issues is once again about reaching out, not striking out. And, as David Schindler said about the tar sands, I worry about the dangers of what the "full package" of resource extraction may bring to the Arctic as a whole.

We also need to remember that we Inuit have been fighting for our *right to be cold*. We have been trying to draw the world's attention to the devastation caused by human-produced or human-accelerated climate change. And we have been trying to educate the world that the vast majority of greenhouse gases that affect the Arctic are produced well outside of it, largely by the United States and China. We have been able to argue that we Inuit are victims of by-products created by industries and economies from which we've had no benefit whatsoever. In other words, just a few years ago, we stood solidly together on high moral ground to defend a way of life. Yet our pursuing resource-extraction industries now means that that high ground is fracturing as quickly as the ice is melting. Indeed, pursuing extractive industries that go against some foundational resolutions of the ICC would weaken the council's ability to take strong stances internationally. Aside from our climate change and POPs work, the ICC continues to actively participate in the meetings of the Convention on Biological Diversity and in the International Union for Conservation of Nature, where we bring our Inuit voice to the issues facing Arctic animals. And we continue our environmental work with the United Nations. At the 2012 COP 18, ICC chair Aqqaluk

Lynge repeated our request for the United Nations to consider the human rights implications of climate change, and asked the participating nations to integrate Inuit and Indigenous knowledge and monitoring into environmental assessments. The ICC's intervention also asked the United Nations to help Inuit and Aboriginal communities by establishing a Green Climate Fund that would allow developed nations to earmark a percentage of their contributions to the fund for Aboriginal peoples in their countries. The ICC also called for help for our communities in developing renewable energy options that would aid us in reducing *our own* emissions. Additionally, the ICC requested that global leaders help the Inuit in researching the presence, and the prevention, of black carbon in the Arctic, an airborne component of aerosols that is finding its way to the North.

Given the ICC's continued push for climate change accountability, how will Inuit support for extraction industries in the North affect its international demands? Would the ICC have to change our very mandate, one that has brought us great environmental accomplishments in the past? What's more, we Inuit are people whose lives and culture depend on ice and snow, and yet expanding resource extraction moves us away from efforts to build a sustainable way of life that will reduce greenhouse gases and environmental degradation. As the rest of the world tries to wean itself off non-renewable sources of energy, we would be encouraging and benefiting from the use of these unsustainable technologies. I find it ironic that the Arctic, a region so negatively affected by high levels of CO_2 in the atmosphere, would now be looking to fossil fuels that lie beneath the ice as an opportunity to pull ourselves out of poverty.

Sadly, the cracks in the moral high ground of the Inuit world have been forming for a number of years. In December 2009, at COP 15 in Copenhagen, assisted by Rich Powell, I gave a number of talks about climate change and the Arctic. After one of these, Patricia Bell, the CBC reporter from Iqaluit, came up to me with a tape recorder in hand. She asked me if I would comment on an interview she'd just had with an ICC Alaska elected official who had said that Inuit were now involved in resource development and that they should be exempt from emission mitigation. In addition to this, I learned that the government of Greenland was suggesting this same rationale. Frankly, I was dismayed and told the reporter I would reflect and get back to her. I didn't want to give a knee-jerk reaction to something as important as this. Over the years, at many of these global COP meetings, I had witnessed how difficult it was for the world to come to common ground on issues like emission reductions when countries such as China, India, and Brazil argued that they should be exempt in order to develop, just as America and other European countries had developed. I thought, *Good God, has it come to this? As Inuit, will we now be like the other countries that have hindered the success of a global consensus on emissions?* During a talk I later gave on Indigenous Peoples' Day at the COP, I reminded the audience, many of whom were Indigenous, of how we the Indigenous peoples of the world, the human faces of climate change, had proven to be invaluable advocates in the climate debate. I stressed the need to maintain our own moral compass by relying on the ancient principles that have sustained us for millennia. I told them our strength stemmed from our ingenuity in knowing how to live sustainably, and that knowledge could serve as a model for all nations, compelling the world to make the strong cuts in emissions needed to mitigate climate change. I added that we must resist the urge to compromise those very values by adopting quick fixes to our economic and social problems.

Our influence, I said, springs from our ethical authority, and if we lose that moral high ground, we lose our influence. I urged our own people and our leaders not to lose sight of the larger picture as we faced difficult choices in our hunger for economic development. I ended by saying, "We have come so far, and whatever happens here in Copenhagen, our Indigenous voice must remain strong and united around the world to continue to model and lead with life-centred sustainability as our focus and commitment."

The debate about resource extraction continues. But many in our community are pushing back against the resource-based industries. In 2011 the community of Clyde River first heard about proposed seismic testing in the waters of Baffin Bay and Davis Strait. The air guns used for this kind of oil and gas exploration are known to wound or even kill narwhals, bowhead whales, walruses, and seals. In 2010, under the leadership of Okalik Eegeesiak and her team, the Qikiqtani Inuit Association halted seismic testing in Lancaster Sound; in the hopes of achieving the same results, the community of Clyde River and the Nammautaq Hunters and Trappers Organization delivered a petition of protest to the National Energy Board (NEB) and passed a number of joint motions opposing the testing. They were joined in their opposition to the geological survey by all the mayors of the Baffin Island region, the Qikiqtani Inuit Association, and Nunavut Tunngavik Inc. In June 2014, however, ignoring the Inuit demands that the seismic testing not be approved until an environmental assessment was complete, the NEB gave the green light for testing to start the following summer. Kangiqtugaapingmiut (residents of Clyde River) held a peaceful march and protest, and shared their outrage with the media. In the summer of 2014, Jerry Natanine, mayor of the hamlet of Clyde River, and the Nammautaq Hunters and Trappers Organization launched a lawsuit in hopes of stopping the testing. It seemed to be the only way to protect their hunting grounds, the Arctic environment, and a way of life.

Even those communities that are pursuing extraction business are aware that they may be bringing more environmental damage their way. The Nunavut Planning Commission recently released a report that calls on Nunavut communities to build their own capacity to deal with oil spills that may result from offshore oil and gas development along the coast of Baffin Bay. Many communities are also exploring alternative energy sources, including ecologically sensitive hydro, wind, and biomass technologies. According to Sustainability Co-ordinator Robyn Campbell, in 2014 Iqaluit City Council unanimously approved a Sustainable Community Plan. Over seven hundred residents weighed in on our sustainable future, producing a crowd-sourced document that is threaded with *Inuit Qaujimajatuqangit*—the knowledge and practice of Inuit ways—and the recognition that everything is related to everything else. The plan is also informed by over three hundred studies and reports written in the past ten years.

The Iqaluit Sustainable Community Plan addresses climate change in three ways. It includes data on what scientists predict we can expect in the future. It also honors observations by Inuit elders about our changing climate, recognizing the significant knowledge and wisdom that is gained through the accumulation of daily experience with sea ice, snow, weather, seasons, lakes and rivers, and wind. Lastly, the action plan identifies objectives, as well as hundreds of concrete actions that address the climate change impacts we face. Placing elders' knowledge together with scientific data and community action creates an alignment of interests, a shared consciousness of our biggest challenges, and a clearer focus on a way forward for Iqaluit, Nunavut.

(For a detailed, far-reaching account of our Aboriginal role in promoting and using clean power sources, check out *Aboriginal Power: Clean Energy and the Future of Canada's First Peoples* by Chris Henderson, published by Boston Mills Press.)

But many more questions remain.

Are we Inuit ironically going to be the tipping point, the spot where the balance of power shifts to the extraction industries because everyone wants to drill for our Arctic resources? Is it still possible that we Inuit will, on principle, assert that we *can* and *will* leave these resources underground as a signal to the world that we, the population most negatively affected by globalization on so many levels, have decided to refuse to be that tipping point? Is there hope that the younger generation of our country, including the young people of the Arctic, can turn around or, at the very least, slow down this already fast-moving development and call for a more inclusive process that will include all voices, especially those at the grassroots?

But if our future Inuit economic well-being is not in the extraction business, where does it lie?

I see the future of Inuit communities in our greatest untapped resource: the natural talents of our people, especially the younger generation. In one generation, our hunters figured out how machinery worked and were able, without training, to fix anything mechanical. Many of our people still possess this ingenuity. You can see it at the hockey arena and other areas of sports. In just one generation, Inuit youth have learned hockey and excelled. Jordin Tootoo, from Rankin Inlet, our pride and joy in the sports world, is certainly not the last Inuk player who will make the pros. Our natural talents and resources can also be seen in many areas of the performing arts, dance, graphic arts, filmmaking, jewelry making, printmaking, and fashion design. In 2001 Igloolik filmmaker Zach Kunuk won the Caméra d'Or in Cannes for *Atanarjuat (The Fast Runner)*, the first feature film ever to be written and acted entirely in Inuktitut. *Atanarjuat* was also Canada's top-grossing film in 2002. Its lead actor, Natar Ungalaaq (also a renowned carver), later won a Genie for his impressive and realistic role in the French-Canadian film *The Necessities of Life*.

Inuit performers have also made a mark in the music world. Susan Aglukark's original blend of Inuit folk songs, pop music, and country music won her three Junos. And the traditional art of throat singing has been brought to the international stage by a number of young Inuit women from Nunavik and Nunavut, including my daughter, Sylvia; Madeleine Allakariallak; Akinisie Sivuarapik; Evie Mark; and Becky Kilabuk. Others, like Elisapie Isaac, Lucie Idlout, Beatrice Deer, and Tanya Tagaq Gillis, have served as ambassadors for Inuit worldwide, sharing their music and winning awards. We are very proud of Inuit women filmmakers Alethea Arnaquq Baril and Stacey Aglok MacDonald, whose work has been viewed at the Toronto International Film Festival. And our carving, printmaking, jewelry making, and fashion design continue to flourish. Many of our carvers work with soapstone and ivory, creating striking pieces of art, but haven't yet established means to market them throughout Canada or worldwide. The sealskin products designed and sewn by our seamstresses are works of beauty and examples of sustainable production, as they use seal by-products. The work of both Karliin Aariak and Meeka Kilabuk of Nunavut has been recognized at fashion shows in Europe showcasing their beautiful sealskin creations. This is culture match at its best.

Like most Inuit women, my mother was an ingenious seamstress, making all our winter clothing and boots, and designing and assembling the large portable canvas tents that we would take on our hunting and fishing trips. I remember watching her spread pieces of canvas all over our living-room floor, cutting and arranging and stitching until she had turned stretches of fabric into a temporary home. It encourages me to see that many of our people have continued to apply these skills and expanded on them. I am constantly blown away by the imagination and beauty of the work being produced by our artists and artisans, as well as by our young busy mothers and fathers, who work full time and still create attractive decorative and practical items in their spare time. There are no doubt many more untapped opportunities in our communities to create small cottage industries like these that provide financial benefits while leaving a small carbon footprint on our already-taxed environment.

These creative endeavors are valuable not just as ways to provide income and financial stability. You cannot under-estimate the power of the arts as a means of change in dispirited communities. There is much "giving birth" to life through these extraordinary works of art that are so grounded in spirit. I recall someone saying, "If you give birth to something every day, whether it is a piece of art, a poem or a writing, music, dance, jewelry, clothing, or something more practical, you will not want to take your own life." This resonated with me. Having a daughter who was part of the movement to revive throat singing (a remarkable art form that the missionaries had banned over a century ago), I have come to truly appreciate this empowering medium for the younger generation. It links them back to their culture and traditions and gives them a profound sense of who they are. I'm not suggesting performing arts or arts of any kind are the only solution to the feelings of inadequacy and disconnection that seem to plague our Inuit communities these days. But they are one way to ensure that self-worth and confidence in our identity is rebuilt. The arts help us to move out of victimhood and have an immediate positive impact on our youth.

Indeed, as we Inuit move forward, we have to be in the business of not only building institutions, organizations, and industries but also building minds and reclaiming the spirit and human capacity of our communities. We do that when we mix creativity, traditions, and business—in other words, when we achieve a culture match between our industries and institutions and our people. And we have plenty of good examples to follow. Our successful film and television producers Isuma Productions, the Inuit Broadcasting Corporation, and Tagramiut Nipingat Productions are just three examples. The Pirurvik Centre in Iqaluit, headed by Leena Evic, teaches, protects, and promotes Inuit language and culture by working with our elders from the region and, in turn, offers these teachings to younger Inuit and non-Inuit in the local workforce. Pirurvik's remarkable work has included the translation of Microsoft programs into Inuktitut. We also have a number of Inuit-owned airlines in Nunavik and several professional programs, including Akitsiraq Law School in Nunavut. And our self-reliance and economic stability could be increased further by federal investment, not only in resource development but also in communications and banking—services people in the South take for granted but that are sometimes lacking in the Arctic.

Each institution we build or invite into our northern communities must allow us to take control over our lives, incorporating our cultural values, values that are based on a sustainable way of living. We cannot let our culture and traditions be overshadowed by quick means of bringing wealth into the Arctic. We need to regain our autonomy and independence, not build up dependency-producing institutions. But first, and most importantly when looking forward, we must keep in mind that ensuring that we continue to have a frozen Arctic allows us Inuit to *choose* our own future.

Of course, it's not just the big resource companies, or some Inuit groups, that are pushing for a future of resource development in the Arctic. Our own Canadian government, in particular the government of Stephen Harper, has been keen to see Canada become an energy superpower, with the Arctic playing a key role in this strategy. Part of this move has been to champion Canada's sovereignty over the Northwest Passage, including sending armed icebreakers through the passage to claim our stake and patrol the waters.

I feel strongly that Canada does need to assert its sovereignty in Arctic waters and refute the claims of other countries and corporations that the passageways through the circumpolar North are international waters. (Canada's Arctic sovereignty is explicitly supported and referenced in the 1993 Nunavut Land Claims Agreement.) Without Canadian authority over these waters, there will be no control over ships using the shorter route linking industrial areas of Asia, North America, and Europe, and no protection against the potential resulting environmental disaster in northern waterways.

Indeed, many of us are watching the unfolding battles over sovereignty and shipping routes with an eye to the environmental trouble that may lie ahead. As the sea ice melts, our Arctic waterways have attracted international interest from countries and companies that would like to use them as trade routes. Routine international shipping through the Arctic raises the possibility of oil and gas spills and contamination of our pristine and delicate ecosystem. We all remember the devastation of the *Exxon Valdez* spill in Alaska's Prince William Sound, but we don't know how much worse the destruction might be if oil is spilled in waters filled with sea ice.

But the best protection we have is the sea ice, ice that is rapidly melting. And Canada's best way to defend the North is by working co-operatively with other nations to preserve the Arctic environment and ecosystem. Climate change and sovereignty are, therefore, two sides of the same coin.

Yet outside of military intervention, our federal government hasn't shown much interest in protecting our Arctic waterways and our Arctic land. Instead, they've adopted a "use it or lose it" approach. Stephen Harper's government has systematically minimized voices of opposition and done away with environmental impact assessments in order to fast-track development. The December 2012 omnibus budget bill alone made more than 99 percent of Canada's navigable waterways no longer protected, weakened environmental impact assessment requirements, amended the definition of "Aboriginal fishery," and reduced the protection of fish habitats. (As I write, Harper's government has also just objected to the United Nations Declaration on the Rights of Indigenous Peoples, which re-establishes the rights of Indigenous peoples worldwide. Canada was the only member country to reject the proposal.)

I honor the Idle No More movement, founded by three First Nations women and one non-Aboriginal woman from Saskatchewan, and its original intent to confront these destructive policies. In a media interview, co-founder Sylvia McAdam spoke about the Indigenous spirituality and love that have spurred the movement on in a peaceful manner. The foundation of our culture should be driving us in everything we do. Culture is not window dressing—it is based on strong

values and principles. The Idle No More movement's intent
was to empower Aboriginal peoples of this country to, once
again, find our spirit, which is connected to the land, and
which was quelled under colonialism and oppression.

Other great Canadians have addressed our government's
tunnel-vision approach to resource development and the
economy. David Suzuki makes it clear that what we know as
the "economy" is not a living thing, but something we have
developed. He says about the Harper government, "My Prime
Minister regards the economy as our highest priority and
forgets that economics and ecology are derived from the same
Greek word, *oikos*, meaning household or domain. Ecology
is the study of home, while economics is its management.
Ecologists try to define the conditions and principles that
enable a species to survive and flourish. Yet in elevating the
economy above those principles, we seem to think we are
immune to the laws of nature. We have to put the 'eco' back
into economics."

The world's most powerful countries must realize that
our environment, our economies, and our communities
cannot be separated: they are all connected. In the Arctic,
and in countries across the globe, we cannot discuss resource
development without also considering what its human impacts
will be. For years, I have tried to help people understand that
the Inuk hunter falling through the melting ice in the Arctic
is connected to actions in the South, to the cars we drive, the
policies we create, and the disposable world we have become.
So too is that hunter connected to the Small Islander fighting
to save his home from the rising tides on the far side of the
earth.

We must think not only about our economies in human
or environmental terms but also about our environment in

economic terms. In 2010 the Pew Research Center (a non-partisan American think tank) released a remarkable report putting an economic value on the services that our frozen Arctic provides to the rest of the planet. With the white ice reflecting the sun's rays back into space, a frozen land locking away methane gas, and our glaciers keeping water on land, the Arctic acts like an air conditioner for the whole planet. The loss of this ecosystem's services, according to the report, adds up to hundreds of billions of dollars per year, and in the future will total in the trillions.

This calculation makes a strong case against those who have long argued that it's simply too expensive to stop climate change. We can now argue that it's clearly too expensive *not* to act. This kind of thinking, however, is only part of the way forward. Ultimately, it will be a debate over rights, justice, and principles that moves our leaders and our world to courageous action on climate change. Recognizing a unique set of rights held by our entire world will provide that framework. I hope that, in the end, this will help us to reconnect as a shared humanity and protect our shared atmosphere.

The future of Inuit is inextricably tied to the future of the planet. As the Pew Research Center report so dramatically testified, the Arctic is part of the global economy, part of the global society. Because our home is a barometer for what is happening to our planet, if we cannot save the frozen Arctic, can we really hope to save the forests, rivers, and farmlands of other regions? For Inuit, a frozen Arctic allows us to continue to choose our own future, to determine for ourselves how our economy and culture will develop. A frozen Arctic also allows the same opportunity to the rest of the world, as our economies will not be burdened by expenditures of trillions of dollars simply to offset the impacts of a melting Arctic.

In February 2010, the leaders of the G7 countries met in Iqaluit. I spoke at an event organized by Pew, who strategically launched their report *An Initial Estimate of the Cost of Lost Climate Regulation Services Due to Changes in the Arctic Cryosphere* (by Eban Goodstein, an economist at Bard College) at the same time and in the same town as the G7 meetings. I shared my message and encouraged the leaders of the world to find a better way forward, one that seeks opportunity with caution and values what we already have, not merely what we hope for. It would be one that has room for all the people of the world, with the diversity of their cultures, their strengths, and their beauty. I tried to convey that we have to take this principled path that respects economic advancements of our nations and communities, but also protects our shared humanity.

To be the most effective global citizens we can be, to be the sentinels for climate change and the models for sustainable living, we Inuit still have much work to do to educate the world about our Arctic communities. These days, Stephen Harper and his government are putting a spotlight on the Arctic for all the wrong reasons. They want to promote mining and sell business opportunities to the Chinese and other countries, often completely ignoring the fact that the Arctic is a populated land, not empty terrain up for grabs. Other people who are flocking to the circumpolar North are after an outdoor adventure. Cruise ships and travel companies are selling tourists on our wildlife and stunning landscapes. Even when people are focusing on the POPs in our water and marine animals or on the global climate change that is destroying our homeland (not to mention the earth's air conditioner), many still fail to see that there are human beings whose lives and cultures are on the brink of being destroyed. When the vast majority of people think of the Arctic, they still think of polar bears, not people.

Indeed, the public attention paid to polar bears is a good illustration of the way in which Arctic people are misunderstood or ignored by much of the world. In around 2006, a campaign was launched by a number of conservation NGOs, the Center for Biological Diversity, and the Natural Resources Defense Council to add the polar bear to the U.S. Endangered Species Act (ESA). The conservation groups noted that the melting polar sea ice meant that the polar bears' natural habitat was disappearing and that their numbers were declining. Their efforts were largely successful. The polar bear was added to the U.S. endangered species list in 2008, and also, in 2012, to CITES. While the Arctic governments, in particular the Nunavut government, and the Inuit communities were also concerned about the polar bear population and the melting sea ice, they objected strongly to the way in which the international community responded to these threats in the Arctic. The population counts were based on southern scientific research, and didn't incorporate Inuit traditional knowledge. In the past, Inuit estimates of wildlife populations, such as seals and bowhead whales, have differed widely from other scientific estimates. (And recent traditional knowledge research has indicated that the numbers of polar bears haven't declined.) What's more, no one talked to our populations about the effects that an endangered species designation for the polar bear would have on the cultural or economic well-being of our communities. While the ESA would note that "the conservation measures provide that the production, interstate sale, and export of native handicrafts by Alaska natives may continue and that the subsistence harvest of polar bears is not affected," Inuit guides and tourism companies in Nunavut alone earn up to two million dollars annually from American and European polar bear sport hunters. While this sort of hunting wasn't part of our traditional culture, our communities had been conducting it in a measured, responsible fashion by adhering to the quota system put in place by both the territorial and federal governments. Polar bear management in Canada is, in the words of Terry Audla, ITK president at the time of the CITES work, "robust, adaptable, responsible and sustainable." Most importantly, however, once again governments and organizations had moved to impose structures and regulations on Inuit communities without their involvement. Given that 75 percent of the world's polar bears reside in Canada, the fact that Arctic Canadians, predominantly Inuit, had been given no agency in the matter was deeply disturbing to Inuit leaders in Canada, including me.

I didn't take part in the global negotiations led by the international conservation organizations like CITES or the International Union for Conservation of Nature, as our own Inuit polar bear experts attended, including representatives from the Inuvialuit Game Council and the Wildlife Department of Nunavut Tunngavik Inc., as well as the ICC Canada president and others. But an encounter with a couple of conservationists made the disconnect between the conservation organizations and Inuit clear to me.

After leaving ICC and following a talk I had given at a law school at the University of Oregon, two people came up to me and thanked me. They said they knew of my work and respected what I was doing on the issues of climate change. They added that they worked for the Center for Biological Diversity and had been the ones who instigated the petition to put the polar bear on the U.S. endangered species list. They told me that they were surprised that Inuit were angry with them for doing this. They had thought that the listing of the polar bear would add pressure to their own country to lower CO_2 emissions and address climate change.

In many ways, I was relieved to meet these people. A couple of people from Iqaluit had confronted me, saying they felt that my work on climate change had encouraged these kinds of misguided conservation efforts. Of course, I had been working to protect our polar bears—as I had been working to protect *all* ice-dependent arctic life. Above all, that means the people who have lived there for time immemorial, and whose traditionals and culture depend on the ice just as strongly as the seals and polar bears do. Interfering with the Inuit way of life was the very opposite of my objectives.

The first thing I said to the conservationists who'd introduced themselves was, "So you are the people who did this?" and my second question was, "If you knew of my work and the elected role I held all those years, why did you not call me?"

I suggested that they connect with the elected leaders in the Arctic, hear what the communities had to say about their actions first-hand, and perhaps start to bridge some gaps. They were open to this.

I contacted a number of people in our Inuit organizations and governments, as well as some Canadian conservationists. I was hoping that the meeting would remind the conservationists that the missionary approach to "saving us" no longer works in this day and age (if it ever did). That no matter how good one's intentions are, failure to include the people affected in your plans or strategies can make things worse on the ground.

My recent experiences with the POPs treaty and the human rights petition had shown me that reaching out, educating, and changing attitudes was possible. I thought bringing together American conservationists and Inuit leaders could convince the conservationists to change their strategy. In addition, I was hoping that a meeting might help us stop duking it out in the media. Unfortunately, the meeting never came to pass.

A sadly similar scenario unfolded in Europe when animal rights organizations successfully lobbied for a ban on seal products from Canada as part of their protest against the commercial seal hunt in Newfoundland. Although an exemption was included for Inuit sealskin products, the ban makes it more difficult for Inuit to market their beautiful and sustainable sealskin products to Europe—or anywhere else for that matter—because of the negative perception caused by the ban.

Even prior to the controversy over the polar bear and the endangered species list, I had been struck by the tight focus on wildlife instead of human life in the Arctic. At the opening plenary session of the Clinton Global Initiative, in the huge conference room, large screens displayed photo montages about each of the four pillars of the initiative: poverty alleviation, religious tolerance, good governance, and climate change. Along with the rest of the participants, I watched as the first three pillars were illustrated by image after image of families, children, and human faces. I anxiously awaited the pictures that would accompany the fourth area of focus: climate change. I was hoping to see photos of northern families and communities, of Small Island and Inuit children. When the montage moved to climate change, however, the focus of the slide show suddenly shifted. The photographs became entirely impersonal, showing images of droughts, melting glaciers, coastal erosion, and polar bears. There wasn't a human face in sight. The climate change pillar didn't convey the real human impacts already experienced among our communities in the Arctic, or among the residents of the Small Island Developing States, who are witnessing their homes sinking as the Greenland ice sheet melts.

As I sat there, I realized how much was left to do to shift the discussion on climate change to the human impacts, and to reframe the debate among the influential leaders gathered there. There was nothing about this presentation that would lead to anyone's understanding of the right to be cold. The following spring, in May 2006, I assembled my own "commitment," or pledge, to CGI, which was to put a human face on climate change. And I wrote to President Clinton about my concern that the initiative, including the film that opened on their website and the materials they were distributing, still omitted the human impacts of climate change:

> In light of my pledge to the CGI, I feel especially obliged to bring this matter to the attention of you and your staff as you prepare for the next meeting. No doubt you have found in your own work that the images that truly move people to action are those that show the human impacts of global problems. World leaders will understand the urgency of climate change only when they realize that the problem is not some future scenario, but a real human drama, unfolding in the present, affecting the lives of many in vital cultures around the world.

I added, "I would be more than happy to supply your staff with multimedia materials illustrating the human face of the changes on our land and in our communities. I want to ensure that the remarkable collection of leaders who assemble again in September, and all those who visit your website or review your promotional materials, fully realize the impacts."

The Arctic is definitely not the Antarctic.

But I remain hopeful that those from the South will visit the Arctic in a quest for true understanding of its life, including its people, the way two of our former governors general, Adrienne Clarkson and Michaëlle Jean, have done in the past. When Governor General Adrienne Clarkson made her circumpolar tour of Russia, Finland, and Iceland in 2003, the media depicted the trip as a waste of taxpayers' money. Yet I think she was ahead of her time. On the tour were fifty or so other Canadians, four of whom were Inuit (including me; Mary Simon, former ICC president and former ambassador to Denmark; Pita Aatami, president of Makivik Corporation; and Piita Irniq, commissioner for Nunavut), as well as other Aboriginal people from the North, Canadian artists, authors, singers, professors, researchers, and creative thinkers. The emphasis was not on politics or business, but on culture and connection with our circumpolar neighbors. It was about meeting the people of the Arctic and acknowledging the Arctic as a *homeland*, one connecting many nations, not just as a frozen, resource-rich landscape.

I know that when people learn more about the people of the Arctic they will be moved to work with us. When I was teaching at Bowdoin, I was impressed by the students' passionate interest in the Arctic and its Inuit communities. At Mount Allison University in New Brunswick I co-taught with Ian Mauro, who had lived in the Arctic for several short periods and was quite well versed on Inuit issues. After we had spent a number of classes discussing the historical traumas Inuit had experienced, connecting this history to the many problems in our communities, several students started to experience emotional difficulties. It was, apparently, the first time that many of them had heard about this dark history of our own country, and the new information brought many feelings to the surface. Some felt betrayed by their school systems and government, as they were only learning about these issues in their fourth year of university. For some, hearing of our Inuit woundings triggered their own unresolved personal issues. My training and experience as a student counselor came in handy during this time, and we were able to move beyond the initial shock of this new knowledge. Along with further discussions on the issues, I brought in some of my country food—fish and *muttaq*—to share, so the students could have a taste of what I had been describing when talking about our hunting culture. Ian brought in some frozen caribou as well. As everyone, including the vegetarians, ate the delicious Arctic food, one could feel the palpable energy of connectedness in the room. Indeed, eating country food communally, even in Sackville, New Brunswick, had a way of bringing people together in a soulful experience. I was thrilled to share a tiny bit of the Arctic with the students, but I was also moved to see that our Inuit experience, our Inuit culture, held meaning for these young Southerners. Their passion and interest made me hopeful that the more we educate the world about our land and our culture, the more they would join us in the fight to protect it. (The students, under the guidance of Ian Mauro, went on to help with the logistics of streaming one of my public lectures for universities across Canada and other organizations and individuals. The lecture was timed to air during COP 17 meetings in Durban, South Africa, in order to draw attention to what we felt the Canadian and other governments needed to achieve at the convention. I believe the students' efforts to spread the message of Arctic environmental protection was their way of symbolically righting some wrongs of the past, and I truly appreciated their help. In fact, two students, Keleigh Annau and Rachel Gardner, continued to help me as I worked on the early stages of this book.)

But as this attention comes our way, we Inuit have to remain on the moral high ground. We can now argue that the financial costs of failing to address climate change are greater than the costs of changing our reliance on fossil fuels and toxic chemicals, but we have also drawn the world's attention to the ethical dimension of continued climate change. We have asked the world to respect our rights to be able to continue our traditions, our culture, our way of life. We have asked the world to respect our right to an economy strengthened by hunting and fishing. But if we want the world to respond to these ethical imperatives, we have to be able to maintain our own moral compass.

As I was taught so well in my early years working in the Kuujjuaq clinic and later with the Kativik School Board and the Nunavik Education Task Force, the challenges facing us Inuit are complicated and interconnected. They are born out of generations of trauma and injustice that have disabled us from thinking for ourselves the way we have traditionally. Yet everything that we do now must show the same respect for our ancient cultures and our tradition of living sustainably on our land as we are demanding of others. We must resist the urge to compromise our values by adopting the quick fixes to our economic and social problems that the extraction industries and development seem to offer. Instead, we can serve as a model for all communities and nations, using our experience with sustainable living and practices to inspire other communities and nations to make the kinds of strong cuts in emissions that are needed to mitigate climate change. And we can celebrate and capitalize on the extraordinary potential that our culture and our community possess.

Bridging Old and New, North and South

My work with the POPs and climate change has led many to see me as an environmentalist first and foremost, something that I do not consider myself to be. Although I wouldn't deny for a minute that the protection of the environment has been a huge focus of my life and work for the last several decades, I came to this particular mission through the concern I had for our people and my great desire to protect the Inuit way of life. On that front, there is still so much work to be done.

Our Inuit communities have the highest suicide rates in all of North America. Turmoil and pain continue to surround us. Alcohol and drugs have become a popular means by which we attempt to change the quality of our lives. All this has had disastrous consequences—the loss of life and the high socio-economic costs, as well as the loss of personal powers, the loss of wisdom, and the loss of freedom.

My previous work with the students at the Kativik School Board, with the Nunavik Education Task Force, with addiction centers and on political issues surrounding the protection of our way of life has blessed me with a deeper understanding of

where we have come from as a people and how we have moved
from an independent way of life to one that heavily relies on
institutions, substances, and programs. I can see that we have
gone from an unhealthy codependent relationship with the
trading companies, religious missionaries, and governments to
similar relationships with today's governing bureaucracy, social
service agencies, lawyers, and consultants. And our historically
learned distrust of *qallunaat* advisors and institutions has
complicated our relationship with conservationists, animal
rights activists, and environmentalists.

I can also see that we are caught in a vicious circle, that
despite the growth of programs, agencies, and institutions in
our communities, in many areas there is little real evidence of
change. As social problems increase, more social workers are
brought on board; as crime rates increase, more police are hired
and court halls get bigger; as accidents and diseases continue
to increase, more hospital staff are hired—yet many of our
problems do not seem to be getting better. Instead, it is the
institutions that are thriving while our people are becoming
increasingly dependent on their processes and systems. A friend,
originally from the South and now a firefighter and part of an
emergency response team in the North, once said, "Sadly, I am
making a good living as a result of the social problems here."

It's hard not to conclude that the dependency-producing
agencies are contributing to our problems, that they are, in
fact, inadvertently leading many to embrace dependencies
and addictions instead of searching out concrete solutions to
empower our people. I firmly believe that if these systems—
whether school systems, judicial systems, or health systems—
do not contextualize our community's problems, helping
individuals, families, and communities to understand the
historical context from which the problems arise and addressing

their roots in a small way, things simply won't get better. If we don't make empowerment and freedom our goals, we won't move beyond the shackles of control and the answers dictated to us. In addition, if we continue to focus on the small picture, on hard work that isn't producing results, on "headway" that looks good on paper but ignores the ever-bleaker reality, then, as we said in the *Silatunirmut* education report, it's like "being lost but making really good time."

That said, I don't want to give the impression that the Inuit community hasn't achieved some impressive successes. For example, more and more high school students are graduating from our northern schools. Many of our students from across the Arctic have continued on to post-secondary education at universities across Canada and around the world (like the prestigious international schools Columbia and Cambridge), graduating with degrees and finding meaningful work and elected positions in our communities. Many of our Inuit teachers from both Nunavik and Nunavut now have both their bachelor's and master's degrees. And in the North, we are also seeing graduates from our own professional schools: the Akitsiraq law program in Nunavut, designed to increase the number of lawyers in Nunavut and the Canadian Arctic and create a cohort of lawyers who will embrace Inuit values, produced eleven graduates in 2005, ten of whom were women. So we are experiencing successes, but I know that many, many more of our young people have the potential to succeed.

As I pointed out in the last chapter, part of the problem is a lack of culture match between the institutions now in place in the Inuit communities and our values and traditions. It seems that we are always saying yes, accepting the learning arrangements and the living structures from the South and, in essence, replicating a system that is not ours. In the Nunavik

Education Task Force review, we realized how fully we had taken on an education system that was copying the southern system rather than building it from the ground up with the foundation, values, and principles held by Inuit. For the most part, Inuit culture and language have simply been add-ons, taught for brief periods in the week. We need to integrate our traditional wisdom into our schooling and recognize that this knowledge base is wide-reaching and valuable to modern life. Piita Irniq, a former student in Churchill and the second commissioner of Nunavut, now teaches Inuit culture and speaks at a number of universities. After hearing one of his talks, Rosemarie Kuptana, of Sachs Harbour, Northwest Territories, former president of both ITK and ICC International, was inspired to share what resonated with her about her Inuit culture:

> Did you know that Inuit had the knowledge of geometry to build igloos? That Inuit traditionally travelled and navigated by the constellations? That Inuit have intimate knowledge of marine and ocean currents? That Inuit are architects—for instance, no one else has been able to perfect the design of the kayak? That the amautik is the best design for carrying babies long distance? That we have the warmest parkas and footwear anywhere in the world? That young girls and boys did not eat certain parts of the animals because of hormone changes? That Inuit have an intimate knowledge of all living creatures around, including their mating seasons? That Inuit believe everything in the world is inter-related, inter-connected and inter-dependent? That Inuit had designated individuals to heal the sick and deliver babies? That they were our spiritual guides and medicine doctors? That Sila

(Universe) was our God? That in Inuit traditional society, malicious gossip was forbidden to keep peace in the clan? That if you went against these sacred laws you were "eating" the future of your children and grandchildren? That Inukshuks were land cairns that meant life, and they held food and were used as landmarks for travel? That every individual had a role in the family and society? Mine was to read the weather.

As Rosemarie so aptly captures, our traditional knowledge offers a wise and empowering base to build on, and we need to support those who are trying to put this into place in our schools, or there will be no opportunities for our students to realize their potential and improve our Inuit way of life.

The problems in our schools and our homes, our collective challenges and our individual struggles, must be treated as a whole. The problem of alcohol and drugs, for example, cannot be dealt with by focusing only on the alcohol and drugs. We need to heal, build back life skills, regain control of our life and destiny, and experience freedom to make the right choices in our lives if we want to move forward.

I believe that part of this holistic approach is spiritual healing. In the past, some of our elders and leaders were also shamans. Long before the church and organized religion came into the Inuit world, we had spiritual rituals and spirit-building techniques. When the new religions were introduced to us, we left behind many beneficial practices that the new beliefs and ideas couldn't replace. I'm not suggesting that we have to revive *all* our old traditions—there are some that we may well wish not to revisit—but there are certainly other means and ways of healing and building our spirit. When you become spiritually alive, you become more attentive to your emotional and physical well-being. Through healing from our past and strengthening our identity, we can help to change the statistics of suicides in our communities, escape the bonds of addiction, and begin to move forward in self-determination, becoming who we are meant to be as individuals, as families, and as communities.

To look at our problems in a holistic way, we will also have to examine how dependence can be countered by redeveloping a sense of personal freedom. This freedom requires us to understand these issues for what they are. But it also depends on the character skills and motivation that allow us to make wise personal decisions. As I've mentioned, instilling sound judgment and wisdom in our schools and institutions is essential, but as I've pointed out many times in this book, we Inuit have always had a remarkably effective way of preparing our youth for responsible self-direction and inner strength: our traditional hunting culture and connection to the land that so effectively build empowering character skills. It has always been our avenue to personal freedom.

But our hunting culture is dependent on a frozen North. As I said in the previous chapter, we Inuit simply cannot have personal freedom, we cannot have choice, if we don't have the right to be cold, if our homeland and culture are destroyed by climate change.

Before I leave the subject of our hunting culture, I want to make clear that when I talk about the importance of our traditional Inuit hunting culture, I'm not being nostalgic. Our hunting culture is not a fondly remembered relic of the past. It's not history. It's a continuing contemporary way of life. And it's perfectly compatible with the modern world.

In May 2009 I was in Iqaluit delivering the ninth LaFontaine-Baldwin lecture, founded by John Ralston Saul. My talk, entitled "Returning Canada to a Path of Principle:

An Arctic and Inuit Perspective," was being filmed by Iqaluit's Isuma Productions for a later podcast. The person operating the high-tech filming equipment was my grandson's father, Qajaaq Ellsworth. A few days after the lecture, my eleven-year-old grandson and his father went out goose hunting. It wasn't the first time that Qajaaq had taken his son hunting. My grandson has been fishing since he was a little boy, and hunting since he was about nine years old. On one of his early hunting trips, he and his father packed their Ski-Doo and headed off to the land outside of Iqaluit. They were gone for the whole day. When they came back, they were hauling a caribou on the *qamutiik* (sled) behind the snowmobile. In keeping with our tradition of food sharing, my grandson and his father gave portions of the meat to family members, friends, and elders. A year later, my grandson caught his first seal. I was away at the time but was sent a picture of a very proud boy standing next to a seal. Life giving life.

This is the learning our youth receive from traveling on the land and from harvesting the animals. I realize that this experience that our Inuit children still gain isn't much different from my brother Charlie's first hunt, or Elijah's. As they head out on a snowmobile instead of a dogsled, their fathers teach them the same things that Charlie and Elijah learned: navigation to the hunting grounds, how to read the weather, animal behavior and migration routes, how to set up a tent, how to use a gun. But just as importantly, more importantly, in fact, they are being taught patience and perseverance, watchfulness and stillness. They are learning about focus and precision and control. They are being shown how to be calm under pressure, how to be courageous at the right time. They are learning sound judgment and wisdom. And their confidence, pride, and sense of self-worth are being nurtured as they master all this. Inuit teenagers today might spend some of their time playing video games, or soccer for that matter, but many still actively engage in their traditional Inuit culture.

I hope that my grandson's future, and the futures of all Inuit youth, will be found in this positive blend of modernity and tradition, and that we will still have the opportunity to combine the disciplined hunting tradition and character-building skills it develops with the new ways of life. I hope, in other words, that future generations will continue to be global citizens with a vibrant culture that is uniquely Inuit. Yet because of climate change, this future, my grandson's future and the future of generations to come, is melting away.

A GREAT DISCONNECT has grown between our communities, our economies, and our environment. This has resulted in rapid climate change that now spirals out of control and fundamentally threatens our world. Those who have traditionally lived closest to the land, and who today maintain the strongest connections to nature, are now at risk of becoming just a footnote in the history of globalization. As our ice and lands are being lost to melting, rising shores, and severe weather, the traditional knowledge of our lands and environment that has sustained us for millennia has come under threat. Our very cultures are now at risk of melting away. If we allow this historic shift in our world to run its course, those peoples with the best knowledge of our world and those best positioned to defend it will no longer exist as the sentinels of global environmental change. The global community must take another approach.

Indigenous peoples worldwide have been at the forefront in pushing for a binding international environmental protection agreement that is rooted in their wisdom and promotes a broad-based, principled rethinking of our approach to this most important of global issues. As part of that Aboriginal effort, the Inuit and the communities of the Arctic are in a position to be a powerful bridge between North and South, between Western scientists, biologists, and conservationists and Aboriginal traditional knowledge. The process of the hunt teaches our young people to be conservationists in a natural way that complements Western conservation efforts, and our hunting culture is not only relevant for survival on the land—its skills and wisdom are transferable to the modern world. One way of life, in other words, does not have to be at the cost of another and can, in fact, enrich the other. (*Two Ways of Knowing: Merging Science and Traditional Knowledge,* edited by David Barber, with photos by Doug Barber, published by the University of Manitoba, does a great job of exploring this idea.)

But more than just a bridge between cultures and ways of knowing, we Inuit are the ground-truthers of climate change. We are on the front lines of the cataclysmic environmental shifts that are affecting the world, and we have observed and confirmed the changes in the Arctic for decades. We see the local impacts—and help link these observations to global data gathered through satellites or climate models. Our traditional knowledge is holistic but not dated; indeed, it offers cutting-edge insights for science, policy, and citizens worldwide.

I think of us Inuit as the mercury in that barometer and as responsible sentinels of environmental change. We have reached out and continue to reach out to warn the world of the effects of environmental change and to spur people and nations to action.

Our message has resonated with the rest of the world. The UN Human Rights Council has recognized climate change as a human rights issue for all Indigenous peoples. The Office

of the High Commissioner for Human Rights has responded
to our call, and now argues that we must use a human rights
approach in responding to climate change and give those
immediately affected active participation in decision-making.
The wisdom of the land, once heard, strikes a universal cord.

I see this kind of energy and synergy growing everywhere
I go: in the audiences I speak to, the university classes I teach,
and the people I meet in all walks of life. In particular, the
youth from both the North and the South are now embarking
on a path that I find commendable and hopeful. They know
that this is their future, and they are not mired in the politics
of the issues. I am impressed by how they are already seeing,
with clarity, the connections. It is through their fresh eyes and
through their commitment to a better future that I believe we
will see a more sustainable world.

I am confident that we can come together as one on this
planet, that everyone can recognize just how connected we
really are. The world needs to realize that our environments,
our communities all around the world are not separate, and
that our shared atmosphere and oceans, not to mention
our human spirit, connect us all. It is imperative to change
the dialogue about climate change and the environmental
degradation of our planet. It has been my intention with every
talk I give to compel citizens to act and bring about change.
This will require us to move how we conceptualize this issue
from the head to the heart, where all change happens. We need
to reassure society that change will not punish our economies,
but, rather, provide an opportunity to flourish in the future,
creating a better world. I do not doubt that real power lies in
a civil society in which individuals, families, and communities
are aware that climate change is as much about humanity as it
is about industry.

And, finally, for my Inuit community, there is a hugely important lesson to be found in the successes we have achieved in getting the world to pay attention to environmental degradation and to the Arctic. They remind us just how far and how quickly Inuit have come as a people in this new world, and of our extraordinary potential as Northerners, as Inuit, as Canadians, and as global citizens. Our successes also demonstrate that, when we act from our hearts and passions as people, when we act from principle as a nation, putting human rights at the forefront, working from the principle of no harm, we can achieve the sustainability and balance many of us seek with the natural world and with one another.

As I write these words, I am working on plans to relocate once again. In 2013 I made a major move from my Arctic home in Iqaluit to Montreal. With the high cost of living in the Arctic, I could no longer maintain my home on that majestic land with the view of Frobisher Bay. Montreal, where I raised my children for many years, was also a strategically well-placed location to prepare for the travel involved in a book tour. Moving to a familiar city in my home province was good for me on several fronts as I finished up the book. But now, if the housing situation works out, it's time for me to plan a return to my hometown of Kuujjuaq. I intend to keep working on the issues I feel passionate about: as long as I can keep contributing to the dialogue of bettering our communities and our planet, I will do so. But I also want to be able to do for longer periods of time what I've so often missed over the last twenty years or so—pick berries, go fishing, walk the rolling tundra, learn how to sew and bead beautiful things, and spend more time being with and sharing country food with my family and community. I want to—I need to—spend more time on the powerful land where I grew up.

ACKNOWLEDGMENTS

The Right to Be Cold took several years to write, and I am grateful to many people for helping it become what I envisioned.

I am deeply indebted to my late grandmother Jeannie Ipirautaq and my mother, Daisy, who were instrumental in showing me how to overcome obstacles in my life. As single Inuit women, they were true survivors. And I am grateful to Nellie Cournoyea and Mary Simon, both accomplished leaders, for paving the way for Inuit women, including myself, to enter elected politics.

Country food has nourished me spiritually and physically, and I thank my brother Charlie, a senator; my godson, Jonathan Grenier; and Pamela, my late sister's firstborn, and her brother Steven for their generosity in sharing it. While living in Iqaluit, I was fed in spirit, friendship, and with country food by my cherished friends Leena Evic, Eva Aariak, Meeka Kakudluk, Myna Ishulutak, Rosie Naullaq, Karliin Aariak, Jamal Shirley, and Kevin and Jeannie Kullualik.

During my thirteen years in Iqaluit, I was welcomed by everyone and blessed to have helped raise my older grandson, Mister Lee, along with my daughter and his father, Qajaaq Ellsworth. As my grandmother guided me, I hope my influence on my grandson will affect his world view. My time in Iqaluit was special, meaningful, and necessary in order to maintain my grueling schedule.

My friend Sandra Inutiq—a lawyer and the language commissioner for Nunavut—deserves a special mention, as she and her husband, Walter, made me feel like family while I lived in Iqaluit. We shared a love of cooking and were both raised with the wisdom of elders. We had many sisterly discussions about how much more we need to do as a people to address all that ails us. Our common commitment in the protection of our environment kept me grounded throughout my time in Iqaluit.

During our work on the Nunavik Education Task Force, Ken Low of the Action Studies Institute of Calgary was instrumental in showing us how to better understand what we knew in our hearts. He taught us to articulate the challenges we face as a people by relating it to the larger human journey. I thank Ken for those insights and for helping me understand the important linkages between the challenges in our communities and the way in which institutions can compound rather then alleviate those struggles.

I've had many great guides in my spiritual journey. Marianne Williamson started me on this remarkable pursuit with her book *The Return to Love*. The spiritual and emotional guidance I received from my gifted guide Peni was vital for me in order to heal and reclaim the essence of who I was meant to be. She has shown me what I needed to see and feel so I could move beyond my wounds and transform my life.

All the success I achieved in my ICC work would not have happened without the loyal support of my team. Terry Fenge, Stephanie Meakin, and Paul Crowley saw me over tough political and personal hurdles, along with Martin Wagner and Donald Goldberg, who brilliantly prepared the Inuit petition that connected climate change to human rights. I thank the ICC Canada administration and Carole Simon, Gela Pitsiulak, and Mali Coley—my executive assistants when I was ICC Canada president and chair—for their professionalism and support. I thank Rich Powell and Sasha Earnheart Gold for their valuable input and for traveling far to gather testimonies for the petition. Rich stayed on with me, drafting many of my talks and attending important global political events as my assistant, and I am indebted to him for his focus, maturity, calm, and brilliance in making my work so manageable during that turbulent and tiring period.

Thank you so much to my fellow petitioners, the wise hunters, elders, and women from Alaska and Canada.

After leaving the ICC, I would not have been able to continue my life's work, including writing this book, without financial support from the Sophie Prize, an anonymous donor (who sent a generous gift when I did not receive the Nobel Peace Prize), Susan Kaplan of Bowdoin College, and Ian Mauro of Mount Allison University. Ian also raised funds so I could teach with him and write. This book was written with the financial support of Mount Allison, the Walter and Duncan Gordon Foundation, the Salamander Foundation, the Oak Foundation, and the Social Sciences and Humanities Research Council. Students Keleigh Annau and Rachel Gardner edited the first draft of this book and helped me with my research.

I greatly appreciate the expertise of Harriet Keleutak, Sarah Aloupa, and Annie Grenier in helping me make sure the Inuktitut words in this book were written correctly. *Nakurmiik!*

I thank First Air, our Inuit-owned airline, for its continued support.

Last, but not least, I am grateful for the editing team at Penguin Canada, including Nick Garrison and Sandra Tooze. I would also like to thank my agent, Rick Broadhead, and copy editor Marcia Gallego. Notably I thank Meg Masters, a freelance editor who worked her masterful (pun intended) magic to make this the book I had envisioned. I believe that the right people come into your life when needed, and her timing was perfect.

INDEX

SHEILA WATT-CLOUTIER is one of the world's most recognized environmental and human rights advocates. In 2015 she received the Right Livelihood Award, widely considered the "alternative Nobel Prize." She was nominated for the Nobel Peace Prize in 2007 for her advocacy work emphasizing the impact global climate change has on human rights, especially in the Arctic. She has been awarded the Aboriginal Achievement Award, the UN Champion of the Earth Award, and the Norwegian Sophie Prize. She is an officer of the Order of Canada, and from 1995 to 2002 she served as the elected Canadian president of the Inuit Circumpolar Council (ICC); in 2002 she was elected international chair of the council. Under her leadership, the world's first international legal action on climate change was launched with a petition to the Inter-American Commission on Human Rights.

BILL MCKIBBEN is a founder of the grassroots climate campaign 350.org and is the Schumann Distinguished Professor in Residence at Middlebury College in Vermont. He received the Right Livelihood Award in 2014 and is a founding fellow of the Sanders Institute. He has written a dozen books about the environment; his first, *The End of Nature,* was published twenty-five years ago, and his most recent is *Radio Free Vermont.*